SCENES FROM THE PAST

STOCKPO
TIVIOT DALE

A CHESHIRE LINE REMEMBERED AND REVISITED

Tiviot Dale. early 1960's. A surprisingly empty car park enabled Ray Keeley to capture the impressive external elevation fronting Tiviot Dale. The station opened in 1865, was built with continuously arcaded frontage comprising a total of thirty bays, its main features being the Dutch gabled centre section with a projecting seven arched arcade. It is perhaps somewhat ironic that most use was made of the car park after the station had closed in January 1967 at a time when Stockport had not started to come to grips with the problems of car ownership. Arguably the best located station for the developing town centre, Tiviot Dale unfortunately did not have a train service to take advantage of the situation and lack of foresight from all quarters allowed everything to vanish irretrievably. *R. Keeley.*

GREGORY K FOX

Introduction

It is now approaching twenty years since putting pen to paper for the first edition of *Stockport Tiviot Dale - A Cheshire Line Remembered.* A lifelong Stopfordian, my 41 year association with the rail industry was well over half way into its duration. Tiviot Dale station though was where the memories of rail travel really began. From roots in Liverpool, it had taken two generations for our family to arrive first in the Manchester area and then Bredbury. One side however only moved the short distance to Runcorn and it was the weekends beside the Manchester Ship Canal that introduced me in the late 1940's as a toddler to homely Tiviot Dale for the train journey to Farnworth (for Widnes). During these Merseyside sojurns however my dad had taken me to Edgeley Park (January 1952) and it became a passion to follow County, home and away. Their away fixtures invariably were seen as a commercial opportunity by British Railways to put on extra trains and one trip to Derby in the Third Division (North) days in particular stays in my mind. I didn't take the number but the train engine was a 'Jubilee'at the head of nine coaches. On another occasions, a County 'special' was laid on from Tiviot Dale for a 'Cup' game at New Brighton, which included a trip across the water from Liverpool. County drew the match but lost the replay at Edgeley Park! Later, after moving to Romiley which, although well served by buses, wasn't as enjoyable as the seven minute dawdle down to Tiviot Dale, located slap in the middle of the town.

Over the years, first as a writer, then publisher, Stockport and its small network of railway lines has always been in the forefront and there was rarely, in over two decades of Foxline, a publication which didn't include parts of the system in and around the town. The first edition of this book appeared in 1991, a few months before the 25th anniversary of Tiviot Dale's closure. Over the last year or so there have been frequent requests to reprint some of the 'older' titles in the series but in this instance, rather than a straightforward repeat, it was decided to add a few pages with additional material. The temptation has always been there to include 'scenes from the present' but in the case of Tiviot Dale and its close proximity to the M60 (formerly M63), the Highways authority can look after themselves. Portwood viaduct and the coal drops vanished in their entirety to be replaced in 2004 by one of the supermarket 'giants'. The readers will see for themselves the changes that have taken places without the need for me to include illustrations of tree infested landscapes that once 'lived' to satisfy a nation's transport needs. ***Greg Fox - Romiley, September 2008***

First Published by Foxline Publishing 1991
ISBN 978-1-909625-43-3
Reprinted April 2015 by Booklaw Publications, 382 Carlton Hill, Nottingham, NG4 1JA
Printed by The Amadeus Press, Cleckheaton, West Yorkshire, BD19 4TQ

(Right) Stockport, Stewart Street c.1966. Shortly after leaving Georges Road, the line descended at 1 in 69 into two short tunnels, Wellington Road (246yds) and Brownsword (39yds). Both possessed shallow cover, the former being only a few feet beneath the goods yard of that name and the latter beneath Wilkinson Street. This view towards Tiviot Dale shows the rock cutting adjacent to Stewart street with a Stanier Class 5 engine running light, tender first and signalled for the through 'road' through the station.The photographer positioned himself above the eastern entrance of Brownsword tunnel which shortly after, in March 1968, was opened out due to deteriorating condition. Right of way over the tunnel was retained by means of a footbridge recovered from Oldham. The trap point to the right afforded protection to eastbound traffic in the event of runaway vehicles. At one time, a small signal box with the name of Stewart Street cabin was located in the recess adjacent to the signal on the right but is thought to have been abolished when control was transferred to the signal boxes at Georges Road and Tiviot Dale in the early years of the 20th Century. The railway is now but a memory, Stewart Street and the collection of buildings to the right giving way to the motorway (now M60), or A560 by-pass as it was proposed originally. ***John Fairclough***

Historical Summary

The effects of twentieth century transport policy carried out in the nineteen sixties saw the elimination of many miles of railway. The age of the motorway arrived with a vengeance and the countryside quickly became criss crossed with the scars of disused ribbons of land that had once provided the ways and means of transporting the nations needs by rail. The line through Tiviot Dale survived this initial onslaught although not without some sacrifice. This east/west route provided an extremely effective and unobtrusive way of getting traffic through the town, and although not of a particularly local nature, large amounts of freight were kept off the ever increasingly constrained nature of the local road system. For many years there had been plans to construct an inner-bypass as relief for the A560, and it became more than critical when, in the mid 1960's, creation of the Merseyway Shopping Centre across the A560 virtually strangled any chance of an effective route through the town for some time to come. Tiviot Dale Station occupied a unique position on the fringe of the main shopping and commercial areas, much better placed than its neighbour across at Edgeley, but facing stiff competition from other public transport. Stockport had always been well served by the trams and buses, and the frequency with which they served the outlying districts was in direct contrast to the spasmodic nature of the train services, stable but unenterprising to say the least. For passengers journeying east across the Pennines or to the East Midlands, Tiviot Dale was ideal, but not enough to put off that fateful day at the beginning of 1967 when, after just over a century of operation, the station dispatched its last passenger train, a victim of the Beeching Axe.

The opening on **1st December 1865** of the Stockport Timperley and Altrincham Junction Railway's (S.T.& A.R.) station in Teviot Dale gave the town its first east/west through line, providing a link between the Stockport & Woodley Junction Railway (S.& W.J.R.) at Portwood and the Cheshire Midland Railway at Altrincham. The Woodley line from the terminus at Portwood was the first of a group that was ultimately to form the basis of the Cheshire Lines Committee. Having been incorporated by an Act of 15th May 1960, the company was empowered to construct a line from Stockport (Portwood) to Woodley where it connected with the Manchester, Sheffield and Lincolnshire Railway (M.S.& L.R.) extended Hyde Branch. Traffic over the 2 mile 61 chain railway commenced on 12th January 1863.

On the eastern side of Stockport, the ST&AJR, incorporated by an Act of 22nd July 1861, set about completing the $8^{1}/_{2}$ miles that separated the Woodley undertaking and the Cheshire Midlands' Northwich Jct. line. The Stockport end of the project possessed substantial engineering works, but the line duly opened on 1st December 1865 with intermediate stations at Baguley, Northenden and Cheadle.

The somewhat brittle nature of the four organisations - the West Cheshire Railway being the one not already mentioned - resulted in the formation of the Cheshire Lines Railway. In an attempt to gain footholds in the region to the south and west of Manchester, the M.S.& L.R. joined forces with the Great Northern Railway (G.N.R.) to exercise some control on the activities of the four railways, as yet unopened. In essence then, the embryo of the CLC came into being on the 14th June 1862 when a joint committee of the GNR and the MS&LR was established. Formal agreement between the two companies by an Act of 13th July 1863 authorised financial and management arrangements for the completion

Continued over

Tiviot Dale, 30th March 1961. Somewhat confusingly, Tiviot Dale continued into the town at the bottom of Lancashire Hill, forming a junction with Prince's Street and Bridge Street. This view from Bridge Street shows the commencement of Tiviot Dale with the station being less than a hundred yards or so to the right but out of the picture. It is perhaps ironic that the main shopping centre of Merseyway now lies immediately behind where the photographer is standing, a situation that would nowadays suit those travelling in from districts to the east and south east of the town. However, to the right of the picture it is just possible to pick out the sign of the Tiviot Hotel, by this time a Robinsons establishment, but formerly the Station Hotel. Just beneath that sign were the British Railways maroon enamel signs indicating the direction of Stockport Edgeley Station - the arrow pointing towards Prince's Street - and of course Tiviot Dale. The protruding black and white half timbered building on the left is the Kings Head public house, still there to this day but providing a reminder of another of Stockport's once numerous breweries, Clarkes of Reddish, which ceased trading in 1963.

T.S. Parkinson courtesy Stockport Libraries.

of the four lines. Keeping a watchful eye on the proceedings was the Midland Railway, a company more than interested in gaining access to the Manchester area. Progressively the Midland made inroads on a piecemeal nature, enabling ultimately a link which allowed their trains from the Midlands and south to use Manchester Central instead of London Road. The section to affect Stockport, the Manchester South District Railway, authorised 1873, was built from Throstle Nest East Junction to the CLC at Heaton Mersey Junction. As a result, Midland trains began to run via Stockport on 2nd August 1880, and for Tiviot Dale, although having been in existence some fifteen years, it opened up completely new travelling possibilities, the most effective arguably being the circular service operating between Central and London Road via Reddish and Didsbury. Other local services prospered, namely those to Woodley and Marple, the railway enjoying a prolonged period of prosperity in the last decades of the century prior to the development of any alternative public transport.

Inter city traffic was at its peak, with Tiviot Dale enjoying the benefits of services to many areas of the country. Competition on the railways was rife and it is difficult to imagine the Pullman Sleeping Cars, Pullman Parlour Car, Special London Express Dining Saloon Car, and a multitude of other such trains linking the station with Bristol, Liverpool, London St.Pancras, Hull, Grimsby, to name but a few. It could not last, as the railway companies strove to knock minutes off schedules, and Tiviot Dale ultimately became a victim of this Victorian obsession. The Midland's desire to reach Manchester in record time resulted in it investing a sum of two million pounds to construct a direct line between New Mills and Heaton Mersey. The route through Disley tunnel and an expansive stopping place at Cheadle Heath gave Stockport an even more direct service to the south, but at the expense of the town station. The opening of that line in 1902 and the developing tram networks saw to it that Tiviot Dale would never again really provide more than a secondary role. That said, the station still played an important part in Stockport's life, particularly Excursion traffic and the special trains that took thousands of Stopfordians to destinations such as Southport, Chester, Liverpool and New Brighton, the North Wales towns, Isle of Man and Ireland via the Liverpool steamers. Parcels, like today, are big business, and vans provided a free collection and delivery service between the station and warehouses, stores, shops and other places of commerce.

The Cheshire Lines, including Tiviot Dale of course, proudly retained its independence until 1948 when it became part of the all embracing British Railways. It still enjoyed the benefit of London expresses both over former LMS and LNE routes and was as busy as ever with freight. Somehow, the St.Pancras services remained to provide that historical link with the past and it is perhaps fitting that those trains provided the last chapter in a story about Tiviot Dale.

Tiviot Dale. 7th July 1908. One of Tiviot Dale's more memorable moments occurred during the first decade of the twentieth century when the then Prince and Princess of Wales, visited Stockport to open the new Town Hall. Their Royal Highnesses were due to arrive at Tiviot Dale station at 3.20 in the afternoon by special train from Chester, and were accompanied by the Duke and Duchess of Westminster along with other guests forming the house party at Eaton Hall. The view shows the Prince inspecting a Guard-of-Honour provided by men from **E** Company, 6th Battalion (T.F) The Cheshire Regiment, under the command of Captain W. Chalinor. The local flavour was extended by utilizing detachments from Stalybridge, Hyde and Glossop along the route through the streets. According to contemporary newspaper reports, the station was a mass of flowers and palms and the platform was covered in red baize with the pillars draped in red cloth with blue and white bands. As the party entered the forecourt, they were greeted by the cheering of four thousand schoolchildren who had waited for more than an hour on the stands erected in the station square. The Royal Party were driven from the station along Tiviot Dale, Prince's Street (late Heaton Lane), and Wellington Road South. Following the ceremonial at the Town Hall, the procession traversed the same route to enable a departure for Chester at 4.40pm. *Stockport Libraries.*

1. Cheadle. early 1960's. Our journey to Tiviot Dale commences in the fields to the west of Cheadle, a stones throw from the river Mersey. This view, with Cheadle CLC in the distance, shows Stanier "Mogul" 2-6-0 No.**42971** at the head of a train of empty carriages. The photographer was standing on the then recently completed Kingsway bridge, a site adjacent to the Styal line of the LNWR. Over three decades later, only the line on which the train is running remains in use. The siding to the right has long since gone and the void filled in with vegetation. Maturing trees have also enclosed the railway formation, although the rural tranquility of the scene has been transformed by the presence immediately to the right of the picture, of a flying junction connecting the A34 with the M63. *The late T. Lewis.*

2. Cheadle. 6th September 1946. A few yards nearer to Cheadle than the scene above, but almost two decades earlier. A goods train, in the broadest sense, in the ever capable hands of Class J.10 0-6-0 No.**5137** approaches the station environs on its way east. Immediately behind the locomotive are two of the ICI bogie hopper wagons built for the stone traffic emanating from Tunstead although on this occasion likely to be returning after maintenance. In the background are bridge and viaduct carrying the Styal line between Gatley and East Didsbury. The bridge that was to take Kingsway (A.34) over the railway was built adjacent to the rear of the train. The rather overgrown track to the left was designated the Down Refuge Siding, with room for engine, brake van and fifty wagons should the facility be required. There were many examples of signals being sighted on the opposite side of the line they controlled. This example was the "outer home" and operated by lever 4 from Cheadle signal box. *W. Potter.*

4. Cheadle. July 1938. L.N.E.R. Class J.10 No.**5174** builds up speed after its stop at Cheadle with a local westbound passenger train comprising the usual three carriages. At this point the line crosses Micker Brook, a watercourse which to the left empties into the Mersey and, when following its route east through Bramhall, becomes Lady Brook and subsequently Norbury Brook. The bridge, No.66, which carries the railway, was widened in 1907 to accommodate the siding seen here to the right. Once again we see Cheadle station in the distance with smoke stained walls of Manchester Road bridge.
W. Potter.

3. Cheadle. c.1936. A remarkably tree-free scene viewed across the fields to the north of Cheadle station, the embankment carrying Manchester Road beginning its descent to East Didsbury. The signals here provided protection for the station itself, the upper (starting) arm being operated by lever 24, the lower (distant) arm, for Cheadle West, requiring the use of lever 23 to pull it "off".
G.H. Platt.

5. Cheadle. May 1938. This pre-war view at the station depicts a scene which was played out time and time again as train upon train of empty wagons made its way back to the Yorkshire coalfields for replenishment. The signals described above can be seen above the fourth wagon, one of many with the initials or name of its owner. These 0-6-0 locomotives were of Great Central origin, and popularly known as "Pom-Poms". Introduced in 1901 as Class 9.J, they were later given a change of identity by the LNER which christened them J.11's. In 1948 No.**5302** was renumbered to No.**64417**. Note also the signal box nameboard, its serif lettering and background finished in the colours, black and white, of the still independent Cheshire Lines Committee.
W. Potter.

6. Cheadle. 19th August 1961. Summer Saturdays at Cheadle were always busy, but more with passing trains than the stations own service. Excursion traffic produced much business for the railway and residents of Cheadle were quite well served. At Easter for instance there were trips to Southport on Good Friday and Easter Saturday, fare 6/6d ($32^1/_2$p). On Sundays, a day in New Brighton was considered good value at 7/10d (39p) if one included the Ferry Crossing from Liverpool landing stage. For Easter Monday, a special train to Chester and Rhyl was provided at 6/6d and 10/3d (51p) respectively. The view here however shows one of the passing variety, a Stanier Class 8F 2-8-0 No.**48711**, at the time a Northwich based engine, with a train identified as the 9.35am (1M33) Blackpool North to Nottingham Midland. This summer only working was well established and traversed an interesting route which took in the east side of Wigan before picking up the erstwhile Great Central line at Amberswood East Junction. From there, a perhaps somewhat less than exhilarating journey contained via Lowton, Glazebrook and Northenden before brief stops at Cheadle Heath and Chinley. The main reason for the train however, for the residents of Chesterfield, Clay Cross, Alfreton, Pye Bridge, Codnor Park, Langley Mill, Trowell, Stapleford and Sandiacre, Long Eaton and Beeston, was to get them home after their holiday. *E.M. Johnson.*

7. Cheadle, 11th June 1957. The last Down train of the day waits for the signal to depart behind Brunswick (8E) based Stanier Class 3MT 2-6-2T No.**40203**. The shortness of the platforms is clearly illustrated here as three carriages fill the entire length. This Liverpool bound train left Stockport Tiviot Dale at 6.23pm - Saturdays Excepted - calling at all stations to Warrington (arr. 7.06pm) where a three minute pause preceded a very leisurely run of almost one hour for the remaining nineteen miles into Liverpool Central. The last east bound service of the day left Cheadle for Stockport at 6.29pm, this being the 4.40pm ex Liverpool Central calling all stations with the exception of Farnworth (Widnes). The signal box had a 25 lever frame by Stevens of which five were spare. It closed on the 25th November 1965 along with Cheadle West. *R.E. Gee.*

8. Cheadle. C.L.C. c.1962. In the three years prior to withdrawal of local passenger services over the line, Fairburn 2-6-4 tank engine No.**42159** was based at Heaton Mersey. This picture, taken some time during that period, shows her entering the station with an Up train, the 4.56pm from Liverpool Central for Tiviot Dale. The platforms were extremely short, accepting barely three coaches, so on this occasion, some unsuspecting passenger might need to engage in a bout of arm-waving to attract attention to their plight. This view west towards Northenden shows a typically rural station with its goods yard full of coal wagons, a situation remaining unchanged until 7th October 1963 when goods facilities were withdrawn. Passenger trains continued for another thirteen months, being withdrawn from 30th November 1964. Overnight, Cheadle lost its rather pitiful service, a timetable that did not show a train for Stockport until 1.20 in the afternoon, in fact it was much easier to go to Warrington or Liverpool. The hiss of steam has since been replaced by the fizz of another kind, the station having been the subject of a conversion to a pub of that name. *T. Lewis.*

9. Cheadle. 5th June 1963. Our old friend No.**42159** enters the station to make the penultimate stop on its journey from Liverpool. Leaving Central Station at 4.56pm, the train ran non-stop to Hunts Cross, before calling at Hough Green, Widnes Central, Tanhouse Lane and Sankey on its way into Warrington. Padgate and Glazebrook were then served before taking the Godley line at Glazebrook East Junction for a thirty minute all-stations amble into Tiviot Dale. (arrive 6.34pm). 42159 left Heaton Mersey in December 1964, following withdrawal of the CLC passenger service, and saw out its days at Bolton, being finally withdrawn in July 1966. *G. Coltas.*

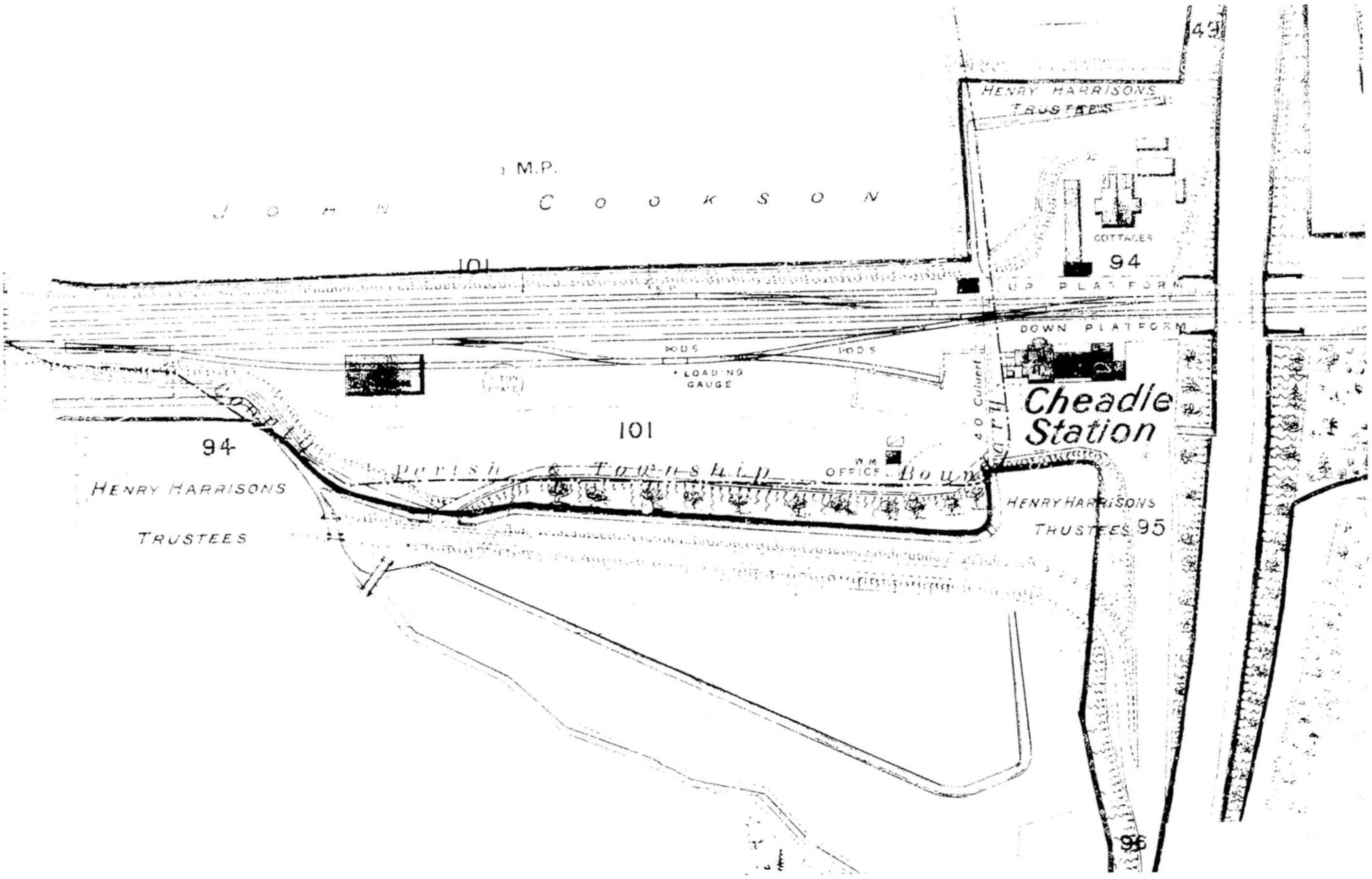

10. Cheadle, 10th June 1950. Few sights and sounds can be guaranteed to stir the soul as a steam locomotive. Stanier Class 8F No.**48503**, although not based there at the time, was to become a long standing resident of Heaton Mersey shed, its final days being concurrent with the demise of the depot. Here she is entering the level confines of the station soon after passing the exchange sidings between Cheadle Junction and Cheadle West. Smoke obscures most of the view through the arch but the train appears to consist of wooden bodied coal wagons. The station has an almost rural charm, with cycles propped against the palings, posters advertising the virtues of going on holiday by rail, etc., Note the use of both oil and gas for platform illumination. *R.E. Gee.*

11. Cheadle West. 1st September 1952. Three quarters of a mile east of Cheadle station were the Exchange Sidings resulting from the building of the Midland Railway's New Mills to Heaton Mersey line of 1902. These extensive sidings were opened on the 1st July of that year and provided interchange facilities between the Midland and CLC companies. The view here shows a Tiviot Dale to Liverpool train, possibly the 3.23pm to Liverpool Central, arrive 4.50pm, passing Cheadle West signal box, hauled by Stanier Class 3 mixed traffic 2-6-2 tank engine No.**40203** of Brunswick (8E) shed. The sidings closed from the 31st January 1971, three years after the ending of steam operation of British Railways, significant because most westbound freight trains from the Midland lines paused here for locomotive crew changes and watering, many of the turns being those detailed in the forthcoming summary of Heaton Mersey enginemen's workings. *courtesy N.E. Preedy.*

12. Cheadle Junction, n.d. Passing Cheadle Junction with an eleven coach special is trafford Park Class 5 4-6-0 No **44717**. The pair of lines to the left of the train came in from Cheadle Heath's North Junction, a section some 32 chains long and more commonly known as the 'Liverpool Curve', opened for goods traffic on 4th May 1902. and to passengers on 1st July of that year. The Exchange Sidings to the left also provided access the siding serving Stockport Corporation Sewerage Works, a location easily identified by those using the more recently constructed Motorway (now M60).The Liverpool Curve was singled in May 1976, allowing simpler working movements and a formation which allowed for construction of the motorway bridge to the rear of the Gorsey Bank housing estate.The roof of Cheadle junction signal box, ultimately destroyed by fire, can be seen above the second carriage. Cheadle west signal box is just visible beyond the rear of the train. Needless to say, the abandoned formation has now been covered by trees. ***T Lewis***

(Above - 12B) West of Stockport, the Mersey valley, apart from the river itself, comprised vast swathes of land on which railways in the Heaton Mersey area were generally the only occupants. Indeed, until closure of some of these lines and the coming of the motorway, vegetation on a large scale was minimal, being more of a pastoral nature. The 1860's changed the nature of things with the coming of the Stockport, Timperley and Altrincham Railway which all but followed the course of the river and can be seen here by the large bridge to the right in 1958. The Midland Railway changed the appearance of things in 1902 with a direct line linking New Mills and Heaton Mersey, at a stroke relegating the route to Manchester via Marple to the title of Manchester Line (Old route). This photograph of 12th July shows the elevated portion of this route at a point where it crossed the east-west CLC on high girders and embankments. An LMS built 4F 0-6-0 No **44101** (17C-Rowsley) heading a goods train of somewhat mixed proportions towards Heaton Mersey. The CLC Heaton Mersey West signal box can be seen beneath the truss girder bridge. **(Below - 12C)** A veteran of 1906, this Great Central 4-6-0 is approaching the end of its life on somewhat mundane duties as it shunts the West sidings of Heaton Mersey during a turn of duty in the autumn of 1947 which has seen it travel from its Gorton base. Still carrying its LNER identification, No **6111**, it had been based some years previously at Heaton Mersey depot Note the tall signal posts which were replaced by those in the view above in 1952. Both photos; ***R Keeley***

13. Cheadle Junction. 17th July 1954. The massive bulk of the River Mersey viaduct provides an impressive setting for another of the former G.C. 2-8-0 locomotives, this time No.**63601**, of Gorton (39A) shed, with a train of empty hopper wagons, no doubt being returned to their home collieries in Yorkshire. Above the engine, a Class 8F rolls down the last few yards of the Liverpool Curve with an ICI stone train bound for Northwich. Stockport Corporation Sewage Works forms the background to a scene since transformed by railway rationalization, only a single line remaining as a reminder. Cheadle Junction signal box to the right of the picture, was the last of the all timber types to survive in the area. It replaced Heaton Mersey West from 23rd November 1969 following minor track and signal alterations in the vicinity of the diesel depot, 29 out of its total of 32 levers being kept in use. *courtesy N.E. Preedy.*

14. Heaton Mersey. 3rd November 1952. Still bearing the original livery of its nationalized owners, this Class J10 0-6-0 No.**65209**, had completed nearly half a century of hard labour since it emerged from Gorton. Its last few years had been spent in the Manchester area, being a Trafford Park engine prior a final spell at Heaton Mersey. The engine is seen here crossing the River Mersey with a westbound train of wooden bodied coal wagons. The bridge, No.72 on the Glazebrook to Godley section, was receiving its periodical coat of paint, the workmen literally perched on equipment that in these days of safety consciousness might at best be considered precarious. The structure had four openings, the outer being skew brick arches with stone facing. The two main spans, of approximately 145 feet each, consisted of wrought iron box girders on the outside with lattice girders of the same material on the inside. A total of four cast iron cylindrical piers, sunk in the river, provided bearing points for main girders. Maintenance was an on going problem and for much of 1963, extensive repair work to the decking resulted in operational upheaval. Occupation of the Down side to Cheadle by the engineers from 19th May 1963 was obtained by sluing and interlacing the Down line into the Up line. This was reversed on the 1st September 1963 for reconstruction work to begin on the other half of the bridge. Everything returned to normal on the 12th January 1964. Following the eventual closure of the route, the bridge was dismantled in 1986. *courtesy N.E. Preedy*

16.(above). **Heaton Mersey West. 19th July 1961.** The familiar lines of CLC pattern all timber signal box. Its 40 lever tappet frame controlled movements of all traffic past the west end of Heaton Mersey's extensive sidings as well as off the Motive Power depot. At the time of its abolition on 23rd November 1969, a total of 32 levers were working with 8 spare. To the left of the picture is the Coaling Stage in the loco depot. The Gorsey Bank housing estate can be seen to the right. *G.Coltas.*

15.(left). **Heaton Mersey. c.1945.** This impressive array of signals stood at the east end of the River Mersey bridge and came under the control of Heaton Mersey West signal box. To gain some impression of size, the posts were 13½" at the base, each one of the four being eight feet apart. The signals, described as "Up homes", controlled forward movement - from left to right - to the Up through sidings, Up Main, Up fast goods and Up slow goods lines respectively. It was replaced, certainly by 1952, by a lattice post signal which carried running line to loop brackets having miniature signal arms. *G.H. Platt.*

17. Heaton Mersey. 24th May 1952. Once again, we see one of the Great Central 2-8-0 locomotives designed by Mr Robinson in charge of a westbound coal train. The engine, Class O4/1 No.**63722**, was on familiar territory, being based at Gorton (39A). The bridge in the background carried the former Midland main line between Heaton Mersey and Cheadle Heath. Constructed in 1901, it was one of the earlier steel bridges with the main lattice girders having a 100 feet span; it was demolished in 1974. The tank engine standing on the Up main will be waiting for the goods train to pass before reversing to the Down side for access to Heaton Mersey loco depot.

18. Heaton Mersey, 15th December 1961. It has been mentioned on numerous occasions about the quantity of coal traffic that passed over the former CLC line. Although nearly all came via Godley from Mottram marshalling yards, relatively small amounts by comparison finished their journeys at the nearby Partington Coaling Basin where stocks would be stored, both in wagons and on tips prior to be loaded into waiting vessels for initial transshipment along the Manchester Ship Canal. One such working was the 9.40am from Godley, seen here passing Heaton Mersey West some fifty minutes after commencing its journey. Former LNWR Class G.2 0-8-0 No.**49426**, a Patricroft (26F) engine, is about to pass beneath the Midland line on its way to Partington Junction where its train of wagons will be transported along the rails of the Manchester Ship Canal's own system by that company's own locomotives. This traffic ceased from 31st December 1966 when Partington (South Side) Junction closed.

G. Coltas.

19. Heaton Mersey, 30th June 1952. This 'birds eye' view from the Midland Railway's intersection bridge looking in the direction of Stockport shows yet another westbound coal train, this time hauled by Class K3 2-6-0 No 61862. These compact engines were first introduced in 1924 for mixed traffic duties and frequently found themselves this side of the Pennines. The engine shed is obscured by the signal box although the coaling stage manages to protrude slightly. The lines fanning out above the engine brought goods traffic in the Down or westbound direction around the back of the sidings. The land both to the left and rear of the signal box is now occupied by the M60 Motorway and an industrial complex respectively. In the distance it is just possible to pick out the arches of Stockport viaduct.

Courtesy ***N E Preedy***

20. Heaton Mersey. 13th April 1965. This morning working from Godley to Northwich was a regular feature of the CLC line through Stockport. Its passage westwards was interrupted at both Woodley and Brinnington Junctions for the pinning down and picking up of wagon brakes respectively. The train is seen here about to pass Heaton Mersey West signal box, hauled by first of the class No.**42700**. Popularly known as "Crabs", these 2-6-0 engines supposedly had an ungainly oscillating movement, hence the nickname. Designed as a mixed traffic locomotive from the outset, a total of 245 were built between 1926 and 1932. Based at Gorton at the time the photograph was taken, No.42700 was eventually withdrawn in 1966 but survived by being taken into the National Collection. Immediately above the engine, a thin line of ballast highlights the formation of the section of line between Heaton Mersey Station and East Junctions. The two lines to the right of the train were the Up fast goods and Up slow goods lines respectively, i.e. eastbound trains would travel forward past East Junction signal box before setting back into the Up sidings, seen here on the extreme right. *G. Coltas*

21. Heaton Mersey, 2nd November 1963. This short headshunt, just sufficient to accommodate a locomotive, was installed during the reconstruction of Mersey Viaduct. It meant that while work on the bridge was progressing, direct access between the main line and the Coal Stage Sidings was not possible. The connection was restored on completion of the work. It is just possible to see the effects of the track alterations in the background.
G. Coltas.

22. Heaton Mersey, 2nd November 1963. The depot was endowed with a large double coaling stage, a split level arrangement in which coal from wagons could be off loaded into small wheeled tubs and deposited directly into the locomotive tender or bunker, whichever was applicable. On this occasion, Jubilee Class 4-6-0 No.**45732** *Sanspareil,* after replenishing its coal stocks, moves off. *G. Coltas.*

23. Heaton Mersey, 1962. Although it has seen its better years, the coaling stage was kept in use until closure of the depot in 1968. The engineers have seen fit to at least keep one half of the structure in reasonable order. The wheeled "tubs" can be seen to the left of the building. The former Midland line from Cheadle Heath runs along the embankment in the background.
R.J. Essery.

HEATON MERSEY

This Stockport shed passed from Cheshire Lines into British Railways London Midland Region Western Division control on 1st January 1948 but retained its former L.N.E.R. locomotive allocation for a while, although the LMS influence manifested itself quite early on. Under the B.R. Motive Power Depot Code allocation, the former Cheshire Lines sheds were designated District Group 13. The Operating District comprised Trafford Park (13A), Belle Vue (13B), Heaton Mersey (13C), Northwich (13D), Brunswick (13E) with sub depots at Warrington, Southport and Widnes, Walton (13F) and Wigan (13G).

The Trafford Park district was abolished in July 1950 with Trafford Park, Heaton Mersey and Northwich placed under Longsight (9A), and were allocated codes 9E, 9F and 9G respectively. Belle Vue was transferred to Newton Heath district and allocated depot code 26G. Brunswick and its sub sheds at Warrington (C.L.) and Widnes (C.L.) came under Edge Hill (8A) District, and were allocated code 8E. Walton (C.L.) and Southport (C.L.) came under Bank Hall and were coded 27E whilst Wigan (C.L.) transferred to Springs Branch control and was coded 10F.

The locomotive allocation at Heaton Mersey in November 1950 (earliest available) was as follows:

LMS			
Class 3	2-6-2T	Nos.	40093,40113/8/24,
Class 3F	0-6-0	No.	43811
Class 4F	0-6-0	Nos.	43836,43945,44080/90, 44144/78,44286,44407/21.
Class 8F	2-8-0	Nos.	48089/99,48127/34/54/55/90, 48208/20/75,48315/6/29, 48406,48503/27/57.48676/82/3.
Class 2F	0-6-0	Nos.	58128,58303.
LNER			
Class D16/3	4-4-0	Nos.	62609.
Class D11/1	4-4-0	Nos.	62663 **'Prince Albert'** 62665 '**Mons**'
Class J39	0-6-0	Nos.	64727/33
Class J10/4	0-6-0	Nos.	65132/5/44/5/6/54/7/60/78.
Class J10/6	0-6-0	Nos.	65181/5/8/93/4/7/8,65200/9.
Class N5	0-6-2T	Nos.	69276,69317/28/31/32/59.

Heaton Mersey had some passenger turns, but the majority of the engine and enginemen's work was freight.

At the commencement of the Summer Schedules in June 1951 there were six passenger engines booked daily, Monday to Saturdays, and two on Sundays. Of these, three Stanier 2-6-2T Class 3 engines worked local services between Cheadle Heath, Stockport Tiviot Dale and Manchester Central.

Loco Turn 23 was allocated to a Class 5 2-6-0 [No.42775] and worked between Godley Junction and Liverpool Central. This train originated from Marylebone at 10/00pm the previous night, (Reporting number 54), and was heavily loaded, made up to 14 vehicles 370 tons and running under Class 'A' headlights. The train divided at Godley with the main part continued to Manchester London Road. The 2-6-0 took over the Liverpool portion, which comprised 1-BCK and 1-BG with another BG from Nottingham and after reaching Central, the loco was worked to Brunswick shed and the Heaton Mersey men travelled home passenger. The loco next worked to Halewood Sidings with a through freight, under the control of Brunswick men Turn 200

Continued over

24. Heaton Mersey. 17th June 1954. This interior view of the shed was recorded to show the recently completed re-roofing exercise, the longitudinal steel frames replacing a more traditional pattern. The engine to the left, a Class 2P 4-4-0 of Midland Railway origin, was a Longsight engine although it was not uncommon to see stock from other depots. The Class 4F 0-6-0 on the right, No.**44407**, had been, and was to remain a Heaton Mersey engine for some years, being particularly well known on the Portwood Shunt and Tiviot Dale Bank Engine turns. *British Railways.*

(MO), or 201 (MX). Meanwhile Heaton Mersey men Turn 24 travelled as passengers from Stockport Tiviot Dale to Halewood on the 7.35am and relieved the Brunswick men. They worked the through freight from Halewood to Dewsnap, returning from there L.E. to Heaton Mersey shed.

The remainder of the passenger work consisted of a Class 'B' passenger working the 6.35am Stockport Tiviot Dale to Risley, due 7.15am with an ex LNER Class D11 4-4-0. The crew then worked as required at Risley until 12/20pm when they were relieved by Heaton Mersey Turn 29 who had travelled out as passenger, and after working to orders until 1/30pm, worked an Empty Coach Stock train to Padgate to form the 1.38pm to Warrington, and worked forward with the 2/10pm to Liverpool Central. They next went on Brunswick shed, where they turned and cleaned the fire before returning to Central to work the 4/40pm to Stockport. The duty on Saturday was different in that the first set of men worked from Stockport to Liverpool at 5.35am, and after turning on Brunswick shed, returned with the same engine and stock with the 8.30am to Manchester Central. The men were relieved by Heaton Mersey Turn 31 who worked back to Liverpool with the stock off the 9.00am from Nottingham, and after turning again on Brunswick shed, the engine and men worked the 4/40pm to Tiviot Dale. The final passenger working was allocated to an LMS Standard Class 2P 4-4-0. According to Richard Strange, of S.R.R.S., ex LMS Standard Class 2P 4-4-0 No.40430 transferred from Chester to cover Turns 33-40. Turn 33 worked the 5.35am Tiviot Dale to Liverpool Central, from there to Warrington C. with the 8.44am, and the 11.05am to Manchester Central weekdays. Turn 35 relieved at Central and worked 12/30pm to Liverpool C., then the 3/55pm to Warrington and returned to Tiviot Dale with the 5/05pm. Turns 36-8 worked the Saturday work starting with the 6.35am to Risley and finished with the 9/00pm from Liverpool C. to Tiviot Dale. On Sunday, Turn 38 worked the 4-4-0 on the 8.50am to Liverpool C., then the 11.30am from there to Manchester Central. The loco was taken onto Trafford Park shed by 9E Turn 104 and the same shed Turn 52 brought it back to Manchester Central and were relieved by Heaton Mersey Turn 40 who worked the 4/50pm to Liverpool Central and the 8/45pm from there to Stockport Tiviot Dale.

Of the remainder of the passenger work, Turns 60/62/63/64, worked mainly with Trafford Park engines between Cheadle Heath, Tiviot Dale and Manchester Central.

The final four passenger workings were Saturdays Only jobs to Blackpool North or Central, relieving crews from Nottingham, Derby or Sheffield at Cheadle Heath. Only Turn 81 worked trains both ways, the other three Turns worked to Blackpool and travelled home as passenger.

Summarising the booked work there were thirteen booked passenger turns on Mondays, fourteen booked turns daily from Tuesdays to Fridays, with seventeen turns on a Saturday and 3 on Sundays. In contrast, there were 92 booked freight turns on a Monday, 101 on Tuesday, 100 on Wednesdays, Thursdays and Fridays, with 97 on Saturdays and 15 on a Sunday, on most days up to ten turns being used on preparation and disposal duties,

25. Heaton Mersey, 17th June 1954. An elevated view from the area behind the engine shed adjacent to the Down goods lines. The new roof is complete and provides a distinct change from the traditional northlight or "saw-tooth" profile. The tall chimney to the right provides a pertinent reminder of the forty or so cotton mills that once existed in the Stockport area, this structure being part of the Gorsey Bank Mill that closed in 1964. *British Railways.*

working to foreman's instructions and working engine movements within shed limits. In addition to these figures, there were between eighteen and twenty control sets daily, which had no booked work but were available on standby to cover extra trains and unforseen circumstances.

It will be seen from the engine allocation list that Stanier Class 8F 2-8-0 was the largest single class and this typified the main work undertaken by the shed, with seventeen engines of this type out of the twenty allocated booked out every week day. Looking at the Engine workings, most of these were handled by Heaton Mersey men, although the 8F's worked as far as Hellifield, Toton and Chesterfield. In several cases men worked one way only and came back 'on the cushions' or alternatively travelled out pasenger and worked the train home. There were no booked lodging turns, and almost all the work done did not qualify for extra mileage payment.

The bulk of the work was local traffic, from the Peak District to Northwich, Warrington, Widnes or the Liverpool area. Most outward freight workings were to Gowhole with nine workings on a Monday. There were seven workings daily to Hartford, Northwich and Rowsley, Five trips daily to Great Rocks and four trips daily to Brunswick and Glazebrook. Trains to Gowhole were mainly empties although there were a couple of workings designated T.F. (Through Freight). Traffic to Hartford were equally divided between E.F. [ClassD] workings and Mineral or Through Freight. Traffic to Northwich was mainly T.F., and that to Rowsley equally divided between T.F. and Empties. All the freight working to Liverpool Brunswick were designated empties, presumably for the C.L.C. Warehouse and traffic to Glazebrook consisted mainly of T.F., Three ex LNER Class J.10 0-6-0 engines worked turn 116. The first loco worked a trip to Macclesfield and return with the same crew. The second loco worked Heaton Mersey Down Sidings Shunt, target No.70, and worked continuously from 6.00am Monday until 2.50am Sunday morning whilst the third J.10 worked Heaton Mersey Up Sidings shunt, target No.71 continuously from 6.05am Monday morning until 6.00am Sunday morning. Another J.10 working Turn 175 worked light engine to Godley at 4.00am and then a pick up goods to Baguley, taking three hours forty minutes for the journey. It returned from Baguley at 9.25am to Heaton Mersey, due 9.40, with another freight, and went on shed for the rest of the day. One N.5 Class 2 0-6-2T engine shunted Godley Down sidings from 12.20am until 11.45 daily (6.00am Sundays), and worked a freight trip to Heaton Mersey before going on shed about 12/35pm. It was replaced by a second N.5 from 11.55am until 5/30pm when it returned to Heaton Mersey West Junction to change over traincrew, returning to Godley Down Sidings again, where it shunted from 6/30pm until 12.10am when it returned to the shed for disposal. Another N.5 worked Target 75 which was the Stockport Tiviot Dale Bank engine. This came out at 6.05am on Monday and assisted until 10.25am when it returned to the shed. It reappeared at 10/40pm and worked as required until 10.25am, traincrews changing over at 5.10am. On Saturdays it continued to work as required until 4/25pm. The final working for the N.5 engine was Target 72 which was the Warehouse Shunt at Georges Road, and came out at 6.10am on Monday and worked daily, shunting until 8/15pm when it worked a trip to Heaton Mersey Down sidings and then to the depot. On Saturdays however, the shunting ceased for the weekend at 1/45pm and the crew worked a trip to Heaton Mersey sidings before working L.E. to the shed.

26. Heaton Mersey M.P.D., 25th February 1968. The end is fast approaching although there is still much activity in the depot. The painted shed code on the smokebox doors illustrates the finality of the situation as locomotives from a variety of locations are sent to their last resting places. To the left, Stanier 8F No.**48197**, long time resident of Toton and recent visitor to Buxton, awaits another turn of duty. She was withdrawn in April 1968 whilst at Heaton Mersey and kept in store there for a few months. The other locomotives on view will no doubt follow a similar path into oblivion. *M.S. Welch.*

32A. Cheadle Heath, 9th June, 1966. As well as providing a variable and intensive range of local train services, the station was arguably second only to Stockport's Edgeley station when it came to inter-city trains. Although the majority of trains from Manchester Central to Sheffield Midland traversed the route via Stockport Tiviot Dale and Marple, this view shows an example of one that provided a somewhat 'semi-fast' service to Hope Valley stations by running express (Manchester Central dep 20.00) to Cheadle Heath (20.13) and thence Chinley (20.29) before an all stations itinerary, including the now closed Sheffield suburban stations of Millhouses & Eccleshall and Heeley - ending at Sheffield Midland (21.38). Stanier Class 4 3/Cylinder 2-6-4 tank engine No 42574 had been allocated to Trafford Park as recently as December 1965 and within a matter of months would be transferred to Wakefield (56A) so this trip to Yorkshire was indirectly preparing it for new home ground, however short lived. ***R Keeley***

32B. Cheadle Heath, January 1966. Some four decades on, a sight like this at Cheadle heath would appear completely surreal. However, for some six years from 1960 onwards, the 'Midland Pullman' provided **the** prestige train service between Manchester and London, albeit Central and St Pancras respectively. Introduced to maintain a premium service betwen the two cities whilst the west coast route out of Euston was being modernised (electrified), the 'Blue Pullman'completed the journey to the capital in 3hrs 10mins, the only intermediate stop being here at Cheadle Heath a short time after the 7.45am departure from Central on its southbound journey. With the introduction in April 1966 of the Mancester Pullman on the newly electrified route, the Midland route, and of course Cheadle Heath, lost the most attractive incumbent of its timetable. ***J W Sutherland***

32C. Cheadle Heath, 23 June, 1962. The activity at Cheadle Heath on occasions defied the nature of its apparently confined space, particularly at the north end where train movements on both the Liverpool and Manchester routes occurred with sometimes alarming regularity. We have here one of those occasions with the platform presence of a Manchester bound local train (left) awaits departure time with Heaton Mersey based Class 4 2-6-4 No **42469** ready for the 'off'. Meanwhile, passage of one of the frequent Tunstead bound hopper trains (empty) from Northwich - off the Liverpool Curve - precludes the passenger train's progress as Stanier 8F 2-8-0 No **48118** (8E-Northwich) crosses the junction at Cheadle North. Both photos; ***R Keeley***

32D.Cheadle Heath, 12th 1958. Cleared for the 'slog' ahead, a climb towards Chinley and the Hope Valley line, this pair of venerable Grimesthorpe (41B-Sheffield) engines work in traditional double headed Midland fashion as they begin there ascent south away from Cheadle Heath. These Johnson designed 0-6-0 freight locomotives, with origins datingback to 1885, were later classified 3F and seen here with No **43243** piloting and **43715** performing the duty of train engine. With a lengthy train of empty coal wagons the engine crews will have testing time for the journey ahead. ***R Keeley***

CHEADLE HEATH

32E (Right) Cheadle Heath, 3rd April 1959. With its heavy commitment to local suburban train services there were extensive carriage servicing sidings to the south of the station. Leaving the sidings with a train of empty stock to form the 7.58am service to Manchester Central is Heaton Mersey's Fowler Class 3MT 2-6-2T No **40067.** Between 7am and 8.31am on weekday mornings, there were no less than six trains which gave Cheadle Heath a direct connection with Manchester Central, four calling all stations with a 22/23 minute schedule and two ex-Buxton workings giving 16/17 minute 'express' scheduling. Cheadle Heath South signal box can be seen through the bridge opening.

R Keeley

32F. Cheadle Heath, 16th December 1966. In the working life of a railway, sixty five years is not an overly extended period of time. That however represented the time, from opening in 1902, until Monday 2nd January 1967, that officialdom, in this case as a result of the 'Beeching' report, saw fit to withdraw a well used amenity. The view here shows the Stockport Road entrance to the station with access to all four platforms via the Booking Hall from which a footbridge ran the width of the station. Cheadle Heath today has but a single line to remind us of past glories. ***Ian R Smith***

33. (Right) Heaton Mersey East Junction c.1963. Had it survived, this box would, today, have been the ideal position to observe the incessant volumes of traffic traversing the M60 Motorway but a few feet away. It was one of the larger signal boxes on the Cheshire Lines system having a 68 lever tappet frame. An all timber structure, its main purpose was to control movements at the point where the manchester South District line came in from Heaton mersey Station Junction, some 66 chains away. Withdrawal of local passenger services between Manchester Central and Derby/Sheffield via Stockport TD/Cheadle Heath, from 2nd January 1967, rendered this short link redundant but it was not closed to traffic until 5th March 1967. The box however continued in use for some time, its life prolonged by a need to back up facilities during a period when track and signalling alterations were affecting the locality. With the ending of steam operation approaching its final stages, the nearby Motive Power Depot had come under scrutiny. It had been recognised that that a limited service facility was going to be needed for diesel locomotives so on 4th march 1968, some two months prior to the demise of the steam shed, a small depot was brought into operational use on a nearby site. Over the next few months, more rationalisation proposals were put in hand, culminating on 19th January 1969, with the closure of east Junction signal box. It had been retained to enable control of the diesel depot entrance/exit to be transferred to the Cheadle Junction box, although as it turned out this was not to happen for some time. The date of closure was to coincide with the virtual ending of of the remaining freight operations at Heaton Mersey. The Up sidings, with a capacity for 588 wagons, were taken out of use, as were the fast and slow goods lines.. The Down slow goods line was truncated and renamed the Diesel Depot line, access being arranged by the laying of a new crossover, operated by hand points, 265 yards from West box. The building seen to the rear of East Junction box and in the centre picture was at one time Faulders Preserve Works. In later years it was use 3d by Cadburys/Frys as a distribution depot, being served by a siding from the Up main line.The ground frame, with two levers, which controlled access to this siding, was released mechanically by lever 52. Upon abolition of the box, control of Faulders Siding was transferred to Georges Road, the ground frame this time being electrically released by lever 44 in the latter box. **(Above-centre) 33A.** Class B1 4-6-0 *Impala* heads a Down freight past the Cadbury/Fry depot at Heaton Mersey East Junction on 24th December 1963. The foreground is now occupied by the M60 Motorway, ***Gordon Coltas***

33B. (Left) Heaton Mersey Down Sidings, c.1950's. Situated on the north side of the Mersey 'valley', Heaton Mersey Sidings stretched for over half a mile between Didsbury Road bridge and the erstwhile Midland Railway route at Cheadle Heath. They were actually segregated into two distinct sites, the Down Sidings at the east end and the Up Sidings at the west end. The view here shows the former, which had a capacity for 423 wagons spread over eleven sidings of different lengths. It is difficult to establish on which line the 'Austerity' loco is standing although the presence of a half hidden engine suggest the main. Closure of this section of Heaton Mersey Sidings from 23rd August 1965 was the the first to affect the CLC route, in fact it was the first large scale abandonment of any railway installation in the town since publication of the 'Beeching Report' (Reshaping of British Railways document of 1962). It was to remain derelict land until the coming of the M63 (later M60) in the late 1980's, when much of this site was swallowed up by motorway and junction. It was also the first site where the integrity of the route was breached with the building of retail units. ***Stockport Libraries***

34. Georges Road. 15th December 1966. The brick built signal box, perched high above the road from which it acquired its name, controlled a busy stretch of railway containing a small number of private sidings as well as the main line between Heaton Mersey East and Tiviot Dale. At abolition in 1983, the box had a 53 lever frame, 16 working with 37 spare. However, this was after the incident which closed the line in Tiviot Dale, and Georges Road had literally become the stub end from Cheadle Junction. Closure of Georges Road to the public in 1967 still left a need for signalling control, particularly with the Cadbury Fry traffic in Faulders Siding - and Clubhouse Sidings which catered for railway engineering needs. Bennetts Iron Foundry Siding, which was a single line alongside the Clubhouse arrangement, had become a private siding by an agreement of 3rd May 1906 but the inevitable decline in usage resulted in termination of that agreement on 30th November 1969. Clubhouse Sidings - the area was bounded by Green Lane, Didsbury Road and Travis Brow - were the next to be reduced in capacity when from the 5th January 1970 six roads were closed to leave 4 remaining. This was the position until final abandonment in 1983. For those interested in movement control to and from Faulders Siding, the small 2-lever frame was released by lever 40 in Georges Road box. A rather more remote operation carried out by Georges Road box was the control of Portwood Stone Terminal. Before the route was severed a 2-lever frame was utilized to gain access to the sidings. This was released by lever 44 in the signal box. Stanier 2-cylinder Class 4 2-6-4T No.**42656** of Trafford Park (9E) pauses between duties outside the box. *G. Coltas.*

35. (Above) **Travis Brow. c.1962.** Passers by at street level could not help notice the signals next to the retaining wall. This trio of signals guarded access from the Up Goods line at Georges Road Junction, the left hand arm for Georges Road yard, centre arm for the main line, and that to the right being for Wellington Road goods. *G.H. Platt.*

36. (Right) **Georges Road. c.1962.** Another example of a Cheshire Lines concrete signal post with miniature arms controlling movements from Georges Road yard. They were replaced towards the end of 1962. All the signalling maintenance was carried out by staff of the Heaton Norris District, which covered an area stretching from Cuddington to Bredbury Junction. *G.H. Platt.*

37. Georges Road Junction. 1962. The signal box above Georges Road, from which it acquired its name, controlled movements within a relatively compact site. On the north side, to the west of the junction, were Clubhouse Sidings, whilst to the east was Georges Road Goods. It was possible for trains to move between the two sites without venturing onto the main line. A similar situation existed on the Down side where the junction for Heaton Mersey Sidings was located. Direct access was also possible from this point to Wellington Road Goods. The view here shows Georges Road box against a background which includes a still busy goods operation and the electrified Manchester to Crewe line passing through Heaton Norris. Stockport's freight requirements were still largely in the hands of the railways, the only real presence on the town's road system, being railway lorries that performed the collection and delivery service around the area. *G. H. Platt.*

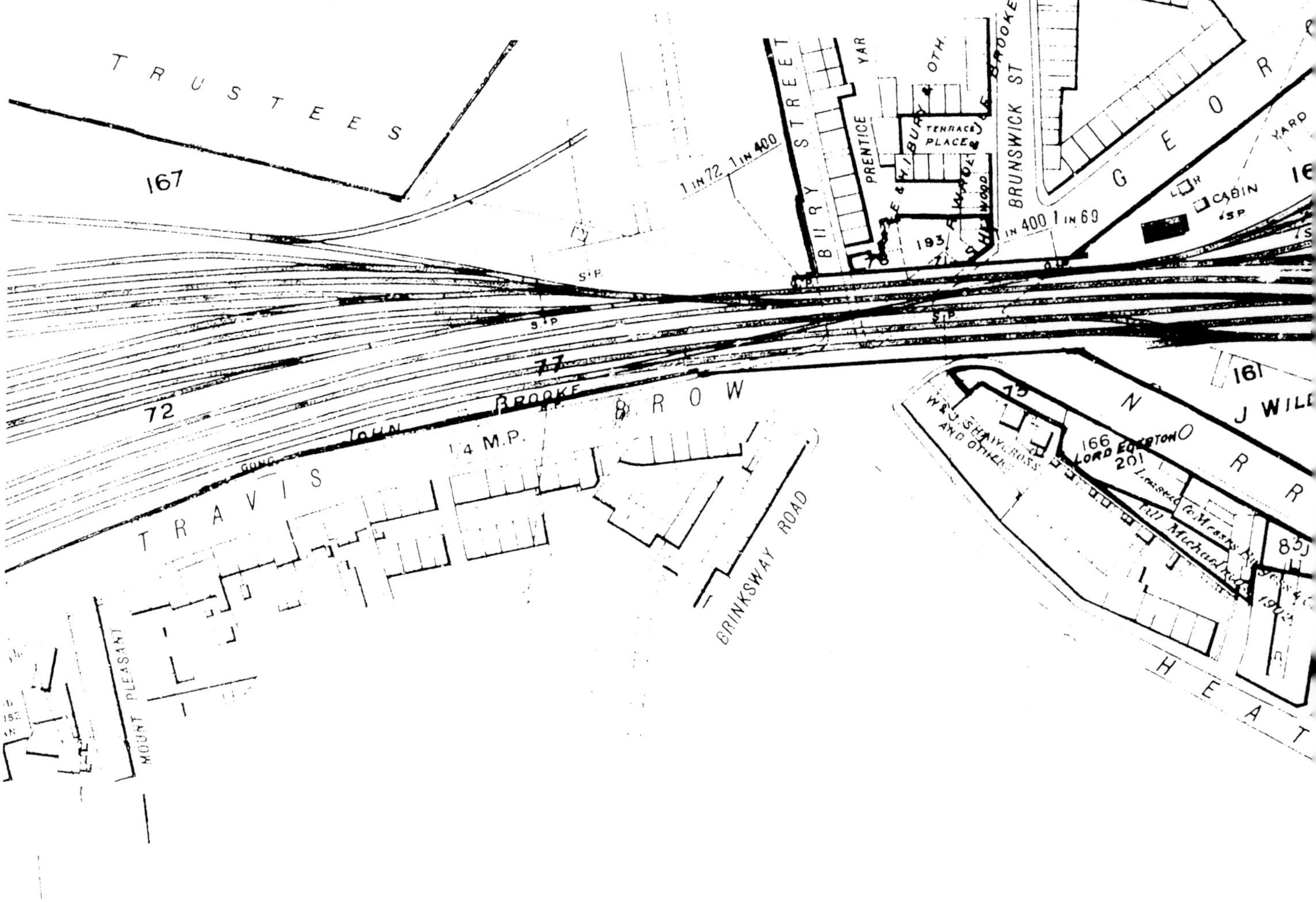

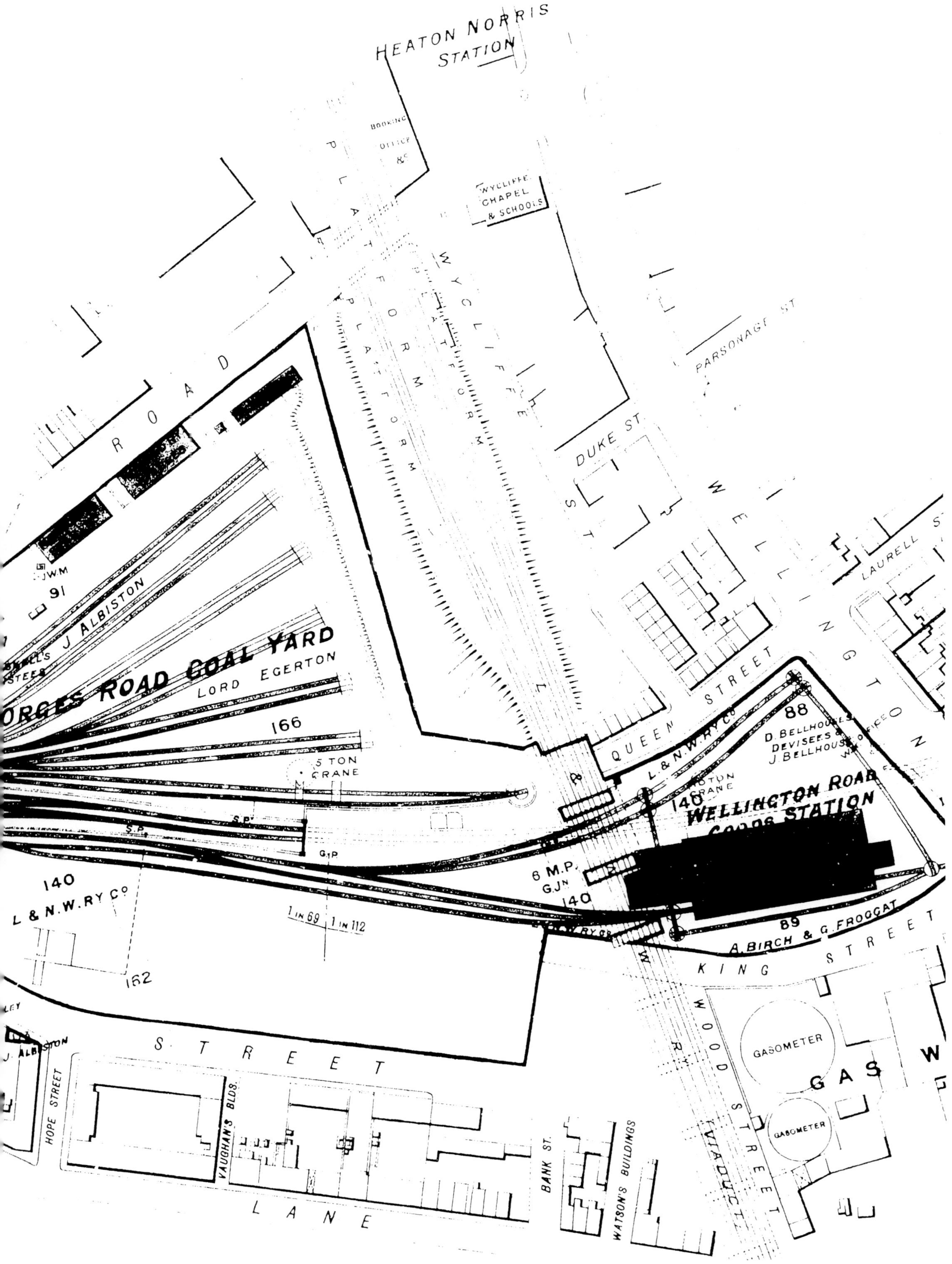

HEATON NORRIS STATION
BOOKING OFFICE &c
PLATFORM
WYCLIFFE CHAPEL & SCHOOLS
WYCLIFFE
PLATFORM
PLATFORM
PARSONAGE ST.
DUKE ST.
ROAD
J.W.M.
91
J. ALBISTON
GEORGES ROAD COAL YARD
LORD EGERTON
166
5 TON CRANE
S.P.
S.P.
G.P.
140
L & N.W. RY. Co.
1 IN 69
1 IN 112
162
6 M.P.
G.J.N.
140
QUEEN STREET
L. & N.W. RY. Co.
88
D. BELLHOUSE'S DEVISEES & J. BELLHOUSE
WELLINGTON ROAD GOODS STATION
140
LAURELL ST.
WELLINGTON
89
A. BIRCH & G. FROGGAT
KING STREET
WOOD STREET
VIADUCT
GASOMETER
GASOMETER
GAS W
J. ALBISTON
STREET
HOPE STREET
VAUGHAN'S BLDS.
BANK ST.
WATSON'S BUILDINGS
LANE

38. Georges Road. 3rd April 1975. The lower end of Georges Road was crossed by the CLC on what can be described as a substantial structure. The formation at this point was widened to take four tracks, a scheme completed in 1884, resulting in these large wrought iron box girders on the outside. This view down Georges Road, looking towards the point where Heaton Lane met Travis Brow, shows the side carrying the main lines. *British Rail.*

39.(centre-left). **Travis Brow. 1989.** Stone tablets displaying the date of construction were placed high in the walls above both Travis Brow and Georges Road. The figures conflict with the official CLC bridge register but presumably the duration of work covered both dates. The example seen here was finally removed in August 1989 when the adjacent site was redeveloped for use as a motor car showroom. *E.M. Johnson.*

40. Georges Road. 3rd April 1975. This elevation on the south side was the view motorists heading for Didsbury Road obtained as they left the town along Heaton Lane. This side of the bridge had carried the goods lines to and from Heaton Mersey Sidings and needed overhead bracing for added strength. Road users today will see odd remnants of the structure amongst the landscaping that has taken place over the last few years. The erstwhile Travis Brow is to the left in a view taken from the corners of Heaward Street and Brinksway Road. *British Rail.*

LONDON MIDLAND AND SCOTTISH RAILWAY.

MIDLAND DIVISION. DISTANCE DIAGRAM. SCALE 1 INCH TO 1 MILE.

MANCHESTER DISTRICT.

SHEET **16.**
(Tenth Edition.)

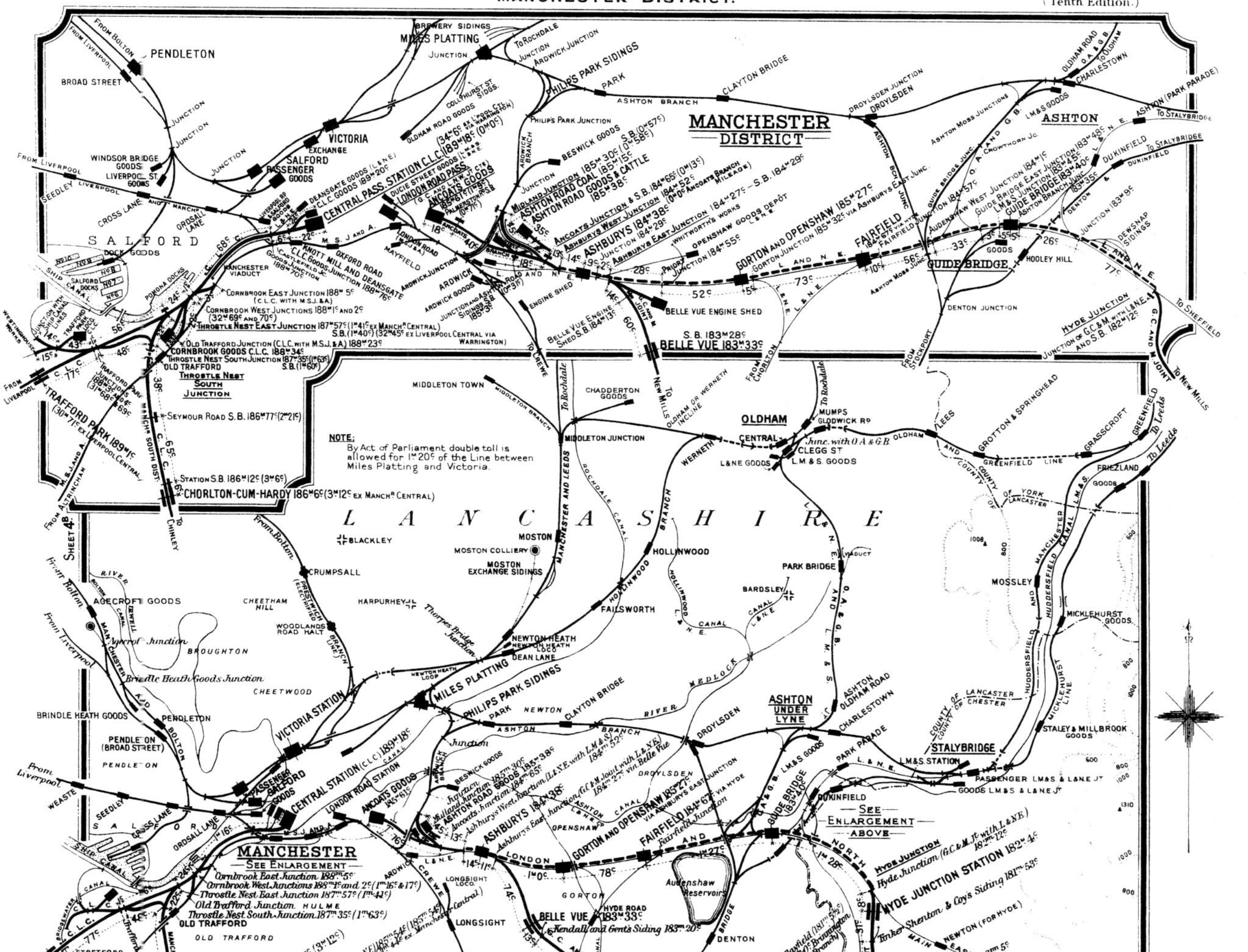

The continuous Distances not in brackets are from St PANCRAS PASSENGER STATION by the Shortest Route. The continuous Distances in brackets represent the Mile Post Mileage.

1923.

EXPLANATION:
MIDLAND DIVISION & JOINT LINES IN WHICH THEY ARE INTERESTED
OTHER L.M. AND S. AND JOINT LINES
FOREIGN LINES
SINGLE LINES ARE SCORED.

(Below) 41. Georges Road, c.1958. There are many views of Stockport that include the town's iconic brick built viaduct but fewer from its lofty parapets. The view here shows looking west is above the sidings serving both Georges Road and Wellington Road Goods. To the left of the photograph are the three lines constituting Wellington Road Sidings which include a spur across the top of Wellington Road tunnel into the goods yard itself. there was a capacity for 122 wagons in the depot. the main line ascended from the tunnel at 1 in 69 to the summit adjacent to Georges Road signal box. At this point the route diverged to form pairs of Goods (left) and main running lines to produce a form of quadrupling as far as Heaton mersey west Junction. The Goods line in the Down or Liverpool direction made its way through Heaton Mersey Sidings. To the bottom right of the picture is Georges Road Depot, at the time in full use although unlikely to be utilising its 124 wagon capacity. Both Wellington Road Goods and Georges Road Depot were closed to the public on 30th September 1967 although the latter continued in use in use as a private siding for scrap metal merchant A Henshaw and Son.In its horse and cart days, Georges Road had three sets of stables, one each for the LNER (MS&LR), LMS (Midland), and Cheshire Lines. The great Northern Company had a facility on the opposite side of Wellington Road in Robert Street. For those both familiar and unfamiliar to the town, this view has been transformed by the construction of the M60 which parallels the route as far as Heaton Mersey West Box that was. Georges Road was transformed into a 'retail park' building over part of and surrounding the entrance to Wellington Road tunnel. The western entrance could be viewed for many years as part of B&Q's garden section although the company later resited on the other side of Georges Road. ***G H Platt***

42.(right) **Georges Road. 18th May 1967.** Passenger services now having gone, the line is left to freight traffic to provide its bread and butter. The main line at this point descended at 1 in 69 towards Wellington Road Tunnel, giving train crews something of a short breather before the climb ahead to Woodley. Here we see British Railways Standard Class 9F 2-10-0 No.**92111** at the head of a train of empty coal wagons from the Liverpool area. The main item of interest however is the bracket signal to the left which protected Georges Road Junction. In Cheshire Lines days, the bracket possessed two of the larger signal arms of the type surviving here. The left hand arm controlled movements to the goods line, right hand the main line. Beneath the main (Home) signal however, is a Calling-on arm which would sometimes be lowered as an indication to the driver of a train that the Starting signal at Georges Road was at danger but that he could proceed in order to clear his train of the adverse gradient present at this point. This did not absolve the driver of any blame should an incident arise, but illustrates the flexibility that had to be occasionally built into an organization that was very strictly governed by its rule book. *G. Coltas.*

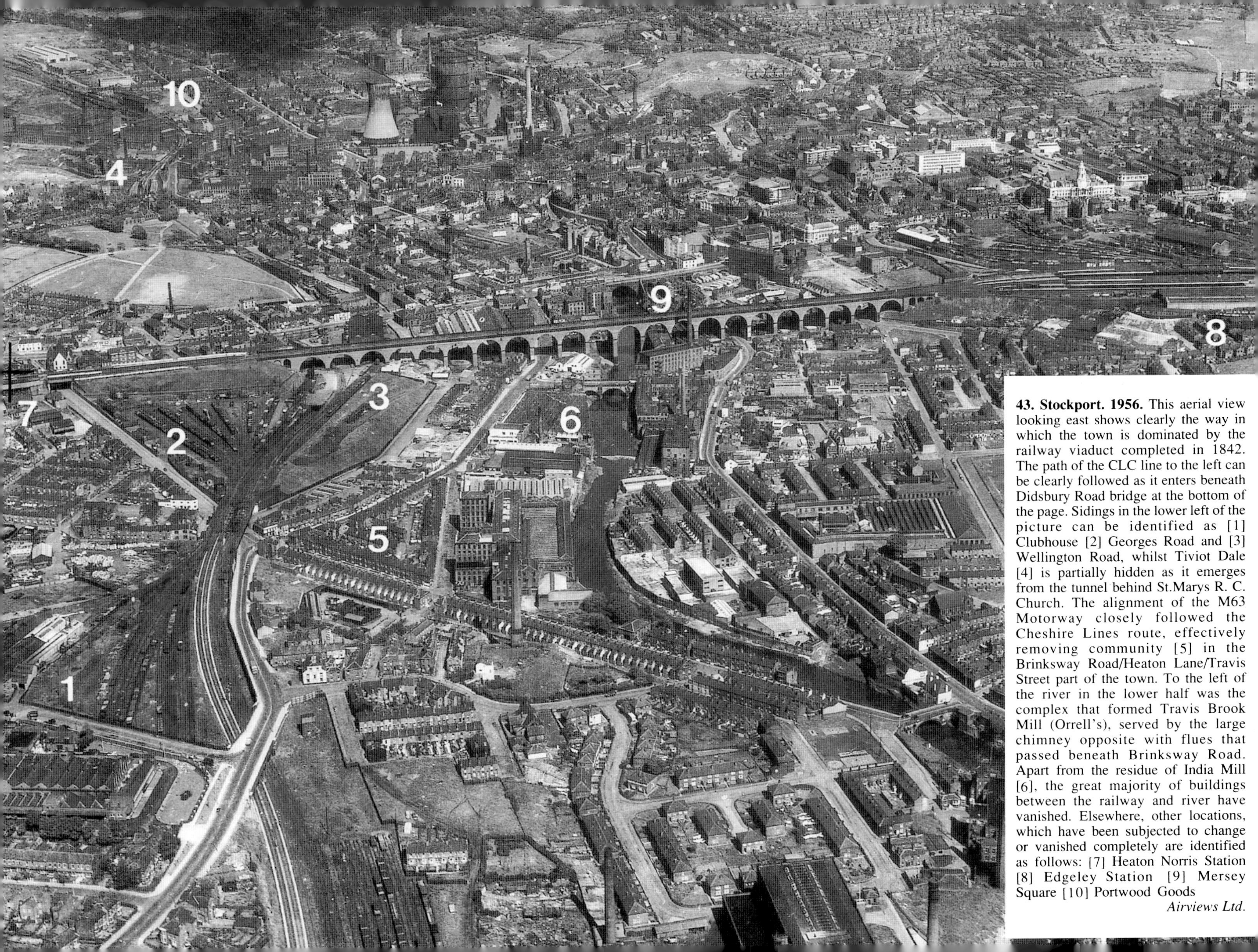

43. Stockport. 1956. This aerial view looking east shows clearly the way in which the town is dominated by the railway viaduct completed in 1842. The path of the CLC line to the left can be clearly followed as it enters beneath Didsbury Road bridge at the bottom of the page. Sidings in the lower left of the picture can be identified as [1] Clubhouse [2] Georges Road and [3] Wellington Road, whilst Tiviot Dale [4] is partially hidden as it emerges from the tunnel behind St.Marys R. C. Church. The alignment of the M63 Motorway closely followed the Cheshire Lines route, effectively removing community [5] in the Brinksway Road/Heaton Lane/Travis Street part of the town. To the left of the river in the lower half was the complex that formed Travis Brook Mill (Orrell's), served by the large chimney opposite with flues that passed beneath Brinksway Road. Apart from the residue of India Mill [6], the great majority of buildings between the railway and river have vanished. Elsewhere, other locations, which have been subjected to change or vanished completely are identified as follows: [7] Heaton Norris Station [8] Edgeley Station [9] Mersey Square [10] Portwood Goods

Airviews Ltd.

44. Wellington Road. 1888. It was business as usual for the Cheshire Lines even though the mammoth task of widening Stockport Viaduct was proceeding quite literally over their heads. Wellington Road Goods Depot consisted of a large three storey warehouse on the east side of the viaduct and was served by sidings passing through the arches from a junction adjacent to Georges Road signal box. This view from the Wellington Road side shows the massive timber centreing required for the construction of the brick arches. A careful look towards the right hand side of the arch will reveal four workmen, only one of whom actually appears to be performing any function. The three wagons to the right are lettered **Cheshire Lines.** the left hand vehicle having wooden or dumb buffers.
British Railways courtesy NRM.

45. Stockport Viaduct. 1888. Work continues apace with the widening work commencing in 1887, construction took place over the next couple of years, finishing in November 1889. As impressive as the viaduct is today, the sight of such an undertaking must have been awe-inspiring for the local population. This view is of course taken from the west side of the viaduct and although not showing any aspect of the CLC, was taken from their land adjacent to Norris Street. Through the left hand arch can be seen buildings belonging to the local Gas Works which occupied the site bounded by King Street (later Bank Street), Wellington Road North and Heaton Lane, prior to it becoming the Tram Depot which opened in 1924. *British Railways*

46. Wellington Road Goods. n.d. Unfortunately this photograph was undated but from the nature of the activity beneath the viaduct, it would be just after the yard had closed to the public in September 1967 when it was adapted for the processing of scrap by Messrs A. Henshaw & Son. The entrance, on the left hand side of Wellington Road North, was virtually opposite to the site now occupied by Debenhams. Wellington Road tunnel was just a few feet below the surface at this point running parallel and just to the right of the goods shed.
Stockport Libraries.

47.(below). **Mersey Square. May 1949.** An unidentified tram car awaits its departure from the Mersey Square terminus of the Stockport Corporation Tramways. This view up Wellington Road North looking towards Manchester shows the Heaton Lane Tram Depot on the left with the roof line of Wellington Road Goods Depot to the rear of the clock tower. *D.F. Tee.*

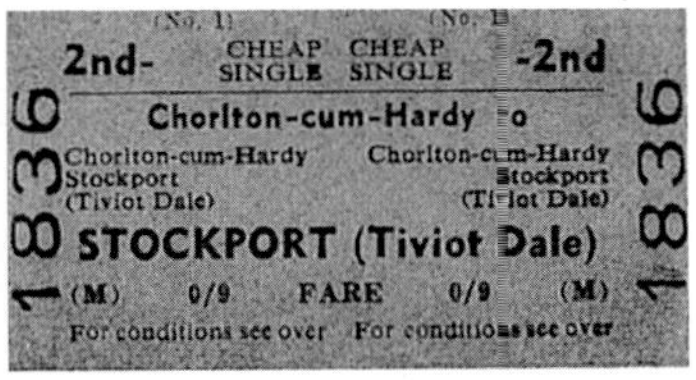

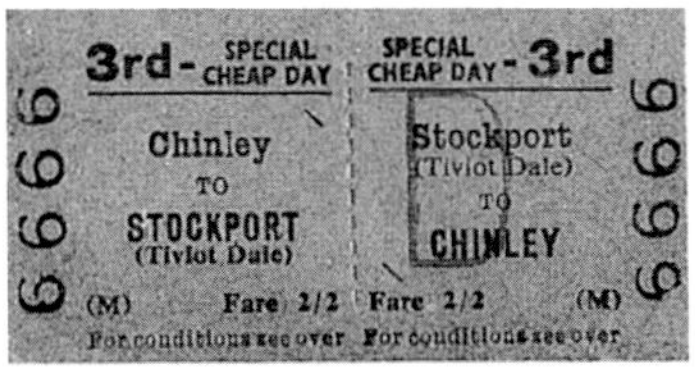

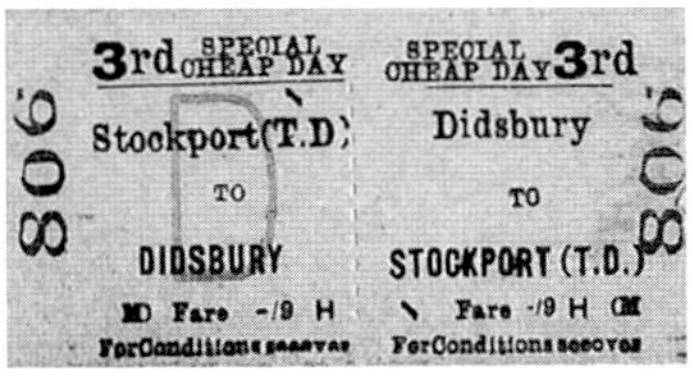

48. Stewart Street, c.1966. The rock cutting behind Stewart Street - now paralleled by the M60 Motorway - is seen a short time before the light engine (see frontispiece) emerged from Brownsword Tunnel. This view east towards Tiviot Dale shows, to the left, the cantilever bracket carrying outer home signals for Tiviot Dale station. The right hand signal is 'off' for the Up Through line. The signal post on the right or Down supported Georges Road distant signal and many years after closure was recovered for use on the East Lancashire Railway. During the days of steam traction, Tiviot Dale tunnel on the rising gradient frequently found this cutting full of smoke from a previous train and it was difficult to during hours of darkness to establish if in fact you were still in one of the tunnels or not. The cutting has been partially filled in since closure with vegetation assisting the site in its return to nature. ***John Fairclough***

49. Tiviot Dale. 1962. A view west from the Up platform at Tiviot Dale. Nothing spectacular, but one of those rare occasions when one could actually see right through. Invariably, the tunnel was full of smoke and all that passing trains did was to move it, somewhat like a shuttle, from end to end. Bored through rock, the 225 yard long tunnel was lined out in brick. The opening at the west end was 25ft 10inches whilst at the east - Lancashire Hill - end, it was wider (31ft 10 inches) as the two lines fanned out into four. Tiviot Dale Tunnel was numbered 80 and the structure abutting it at the east end carrying Lancashire Hill was number 81. This latter bridge had steel main girders, being reconstructed in 1924. To the right of the tunnel entrance was the site formerly occupied by "Tiviot Dale West" signal box, one of two that controlled the station and which were replaced by the all timber building at the east end on the Up side. *G.H. Platt.*

50. Tiviot Dale, 21st May 1965. Over the years, this late afternoon train from Manchester Central to Chinley/Sheffield had become the property of a "Compound" locomotive, indeed train spotters in the Bredbury/Romiley area, of whom the author was one, treated the daily visit with disdain as the same class of engine, more often than not the same engine, carried out this humble duty. It was more a case of "before the Lord Mayor's Show", expectations being high for the appearance of a "Jubilee" class locomotive on the Liverpool to Nottingham express which would follow some twenty minutes later. The fact that it was *Hong Kong* or *Seychelles* for the umpteenth time did not matter, it was **the** train of the day. However, by the early 1960's "foreign" engines had infiltrated and on this occasion, former LNER B.1. Class 4-6-0 No.**61394** enters Tiviot Dale with the 4.05pm ex Manchester Central to Sheffield via Romiley train, stopping at all stations. As well as providing a leisurely journey across the Pennines(116 minutes during the week, 122 on Saturdays), it provided connections for intermediate stations at Chinley for the St.Pancras express. This particular locomotive had quite recently been in auspicious company, having shared billeting at King's Cross (34A) shed until December 1962. From then on it was something of an anti-climax, Canklow, Mexborough, Langwith, prior to eventual withdrawal towards the end of 1965. *G. Coltas.*

51. Tiviot Dale. 1962. Not quite an engineman's eye view of the station but certainly one which took in almost all it had to offer. This view east from the Up platform shows the fairly basic facilities, both sides being linked with an attractive wrought iron arched latticed girder footbridge with a span just in excess of sixty feet. The structure was completely boxed in and although not possessing the action of a trampoline, certainly put a spring in one's step when running for a train. Removal of redundant assets following closure of Tiviot Dale signal box involved uprooting signal posts, lifting track, etc. It was decided to utilize the same engineering possession for the removal of a signal gantry at Portwood as that for demolition of the footbridge and so was arranged for the 17th december 1967. One end of the gantry was built into the Warehouse wall at Portwood and whilst attempting its removal, engineers noticed a bulge in the masonry. As a consequence, work on demolition was held over until the 14th January 1968. The footbridge however, did not escape its fate.

G.H. Platt.

52. Tiviot Dale. May 1961. It is the middle of the afternoon and the crew of LMS built 4F 0-6-0 No.**44407** will be looking forward to booking off at Heaton Mersey in a few minutes time. The engine, working light from Tiviot Dale, had spent the last period of its turn of duty as banking engine for trains requiring assistance towards Woodley. The number 74 above the tender coupling identifies the trip or "Target" under which the working was carried out. From Mondays to Fridays the engine left Heaton Mersey at 8 'o 'clock in the morning to perform its duties as the "Portwood Shunt"; on Saturdays it left the depot at 7.30. Tiviot Dale, forever grimy, was still a busy place, although it saved its operational "peak" until the evening, a fact highlighted in a letter to the Stockport Advertiser on the 29th June 1961 by Mr Brady of Buxton. It is reproduced here and reflects the feelings of many.

H. Challoner

Tiviot Dale station lack of imagination

Sir,—It is refreshing to see the improvements in train service to and from Stockport Edgeley resulting from the electrification of this line, as opposed to the growing number of curtailments and closures we are accustomed to seeing announced in the press and elsewhere nowadays

One is obviously not going to undergo a tiring and hazardous journey into Manchester only to spend a quarter of an hour on getting there to find a parking place, when one can hop onto one of the now frequent trains, relax and reach town in only ten minutes—although I rather wonder whether people who are not actually in very close proximity to the station will be so encouraged to use the new service with the present acute shortage of parking space there as well, and the absence of a bus service up the station approach.

However, in contrast to this, a remarkable lack of imagination is shewn with regard to Stockport's other station—the one nearest to the present shopping centre, the proposed Merseyway development and a good deal of Stockport's office zone—Tiviot Dale.

Tiviot Dale is the answer for the person who at present takes more time reaching Edgeley Station than for the rest of the journey into Manchester by rail; the answer to the crowded roads and buses in the mornings and evenings westwards to and from Altrincham and Cheadle and eastwards to and from Bredbury Woodley and Hyde and the answer for the many people resident in Chapel-en-le-Frith, Chinley, New Mills and district, Marple and Romiley who like—or would like—to shop or work in Stockport without driving their cars through the ever increasing traffic.

What is the present service? Nothing to Altrincham, nothing to Hyde; from Chinley, New Mills, Marple, etc. we have only four trains a day in the morning and afternoon to Tiviot Dale (although there are another five trains between 6-24 p.m. and 11-16 p.m.!! Of these the 10-03 a.m. arrival waits at Stockport for half an hour (offering a through timing from Chinley to Manchester of almost an hour and a half); the 12-56(?) arrival is more often than not anything up to 20 minutes late; the 3-27 p.m. arrival goes no further; and only three trains in the day out of the total of nine enable the inhabitants of the fast growing and already well-populated town of Chapel to travel to Stockport without either changing and waiting for anything from nine to 13 minutes or walking two and a half miles to the country village junction of Chinley.

This sort of service is hardly an inducement to use the railways and it cannot be a very paying proposition, as it is common to see a large express locomotive hauling a six coach train on this line patronised by only eight passengers. It would appear that the men at the top are purposely setting out to eliminate yet another line of passenger stations, which as those readers who are acquainted with these matters will know, is a policy which is being pursued especially rigorously on the London Midland Region by showing as big a loss as possible on "disfavoured" lines.

Let us hope that Dr. Beeching will move in the right direction. He has not started off very well by threatening fare increases so soon after his appointment. But he must remember that any moron can stop a concern losing money by closing it down; the clever man makes it run at a profit. The two obvious ways to start are by offering an attractive service and by cutting expenses. And the latter can be effected, in part, at least, by dismissing a good percentage of station staff who sit on their seats most of the day doing nothing.—Yours etc.,

C. M. H. BRADY.

Parkfield,
Carlisle Road,
Buxton.

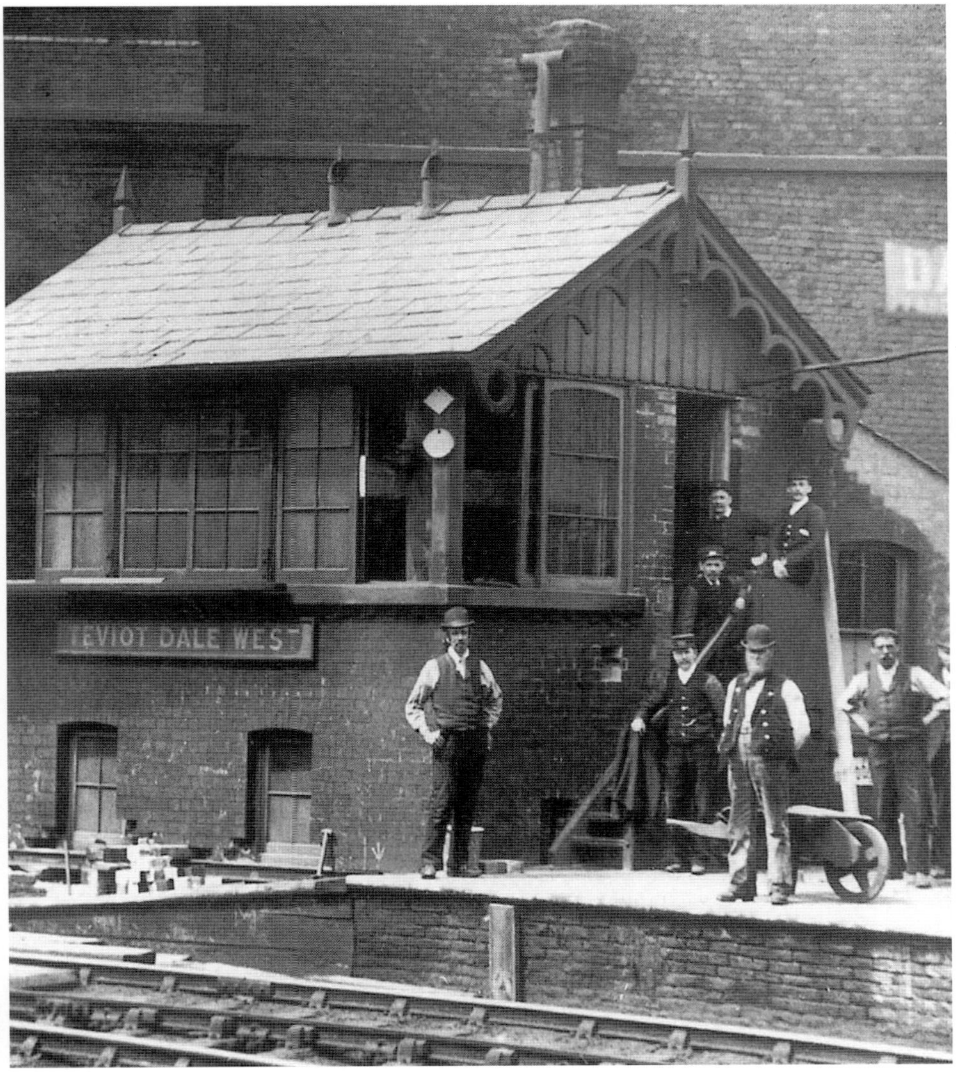

54. Tiviot Dale, 30th December 1966. This bell, located at the east end of the tunnel, is thought to have been rung by station staff upon the approach of a passenger train. In all probability the custom ceased with the closure of west box although the bell itself remained until the station closed. *W.J. Skillern.*

53.(left). **Teviot Dale, c.1901/02.** A pause in the repair work at the station's west signalbox. Situated hard by the Lancashire Hill retaining wall, the box carries the scalloped CLC bargeboard design seen also at Baguley and Northenden. The early spelling 'Teviot Dale' provides one of the few pieces of evidence of the naming of the station in this particular way. On the right hand corner post, there are two small boards, one diamond shaped, the other oval. These were actually "fault" boards, coloured red on one side and, as seen here, white on the other. Displayed thus, 'all right' was indicated; with the colour reversed - a fault was assumed. The diamond shape related to the mechanical side of signalling, the oval one to the electrical/telegraph aspect. The history of this box is a little uncertain although official documentation shows that a Tiviot Dale West box was still in existence until 1931. It has not however been possible to ascertain whether it was this particular building. *Stockport Libraries.*

55. Tiviot Dale, 22nd February 1966. The Up platform buildings that served eastbound trains. Originally, the left hand end of the building provided general waiting room facilities whilst the section to the right contained a refreshment room, toilets for both genders being on the end. *Stockport Libraries.*

56. Tiviot Dale, 31st January 1965. Anyone ascending Lancashire Hill (left) would hardly fail to notice two of Stockport's architectural 'gems', Tiviot Dale Station and Hanover Congregational Chapel. Both were features of the area until the late 1960's and went the same way, in the name of progress. Many will also remember the signs at the junction of Tiviot Dale and Princes Street, which offered directions to both of the town's stations'. Those wanting Tiviot Dale had only a cock stride, those for Edgeley a rude awakening, particularly in wet weather. At the time of the photograph, the author remembers using the train services on offer to travel between Romiley and Wilmslow. Leaving Romiley on the 6.17pm in the evening, a nine minute ride to Stockport was followed by a brisk walk (run) to Edgeley Station where the 6.49pm for Crewe enabled a Wilmslow arrival at 7.01pm. Does public transport offer such a facility almost a half a century later. Readers will also observe that a truncated Lancashire Hill commences further up the hill and a good imagination is needed to visualise Tiviot Dale station above the path of the M60. ***Stockport Libraries***

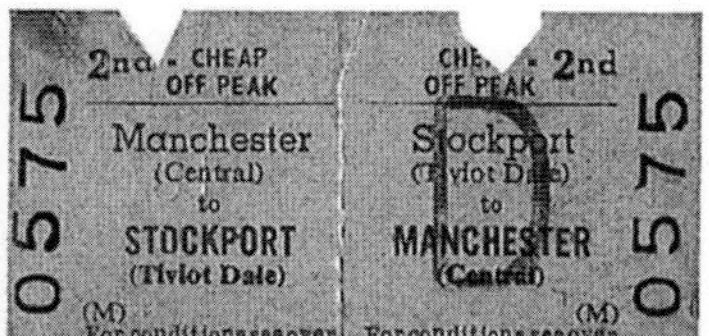

57. Tiviot Dale, 28th December 1966. An almost permanently gloomy atmosphere prevailed beneath the platform canopies. This view along the Down platform - westbound trains - shows an adequate supply of furniture - in the form of the rusticated seating originating from the works of Andrew Handyside in Derby. The bay window effectively extended the Station Masters Office into the platform area, enabling him to view all that was happening. Beyond is the timber framed "Passimeter", the point where tickets were checked and passenger flow controlled as necessary. Steps to the platform opposite were situated between the third seat and bay window. *W.J. Skillern.*

57A. Stockport Tiviot Dale, 29th December, 1966. From the cobbled approach of Tiviot Dale - Tiviot being a corruption of Teviot - we can enjoy the station buildings from another direction, this time the west end of the station. This magnificent facade was into its last few days as an passenger station, more is the pity that such an ornate structure was not preserved for posterity. The site is now several feet above the eastbound lanes of the M60 Motorway. ***G Coltas***

57B. Tiviot dale, 23rd June 1962. The first of two views of an ex-Liverpool Central train alongside the Up platform having just arrived behind BR Standard Class 4 2-6-0 No 76048 with the 4.56pm ex-Liverpool Central. In common with a number of towns which had railway organisations headquarters - Liverpool was home to the CLC - there were trains which provided a fast service home for staff, particularly managers of that company at the end of the working day. This particular service ran non-stop to the Hunts Cross, so satisfying the needs of those residing in the south Liverpool area. There were in fact five through trains on weekdays between Liverpool and Tiviot Dale, a couple of which called only at Widnes and Warrington. The service was withdrawn on 30th November 1964, local stations affected including Cheadle CLC, Northenden, Baguley, West Timperley and Partington. ***G Coltas***

(Above) 58.Tiviot Dale, c.1966. Taken from the site of the former Hanover Church, this view includes more of railway interest. The Down bay platform is in use as carriage sidings but difficult to guess which service as the Warrington/Liverpool trains had been withdrawn some two years earlier. The sidings in the foreground are for departmental use, with both Civil and S&T departments in residence at the station. The concrete unitscomprise drainage channel units and covers, catch pit units and fencing posts, all courtesy of Newton Heath Concrete Depot.

(Below) 58A. Tiviot Dale , 11th April 1964 (Saturday). The full panorama of the station and yard unfolds in this rail level view from the east end of the Down bay platform. Resident at Heaton Mersey Depot (9F) for only a few months, LMS Ivatt designed Class 4MT 2-6-0 No **43031** runs spiritedly through the station along the Up Through line, the route ahead cleared by the lowered signal.To the right of the signal are the remains of the one time locomotive shed, not used for that purpose for many years but still retaining a water tank with which to service the platform water columns. ***J W Sutherland***

59. Tiviot Dale 21st May 1962. There were very few quiet periods at Tiviot Dale, passenger train movements being frequently interrupted by freight operation. In this instance a Down freight will put in a spurt before entering Tiviot Dale tunnel to ensure a successful climb past Georges Road goods were it emerged from Wellington Road tunnel on a 1 in 65 gradient prior to passing Heaton Mersey sidings.Stanier 8F 2-8-0 No **48695** (9F- Heaton Mersey) was on home territory as it takes the Through line. In the background is Class 4F 0-6-0 No **43929**, only recently transferred to Northwich (9G) after many years at Wellingborough (15A). the loco, although appearing to do a spot of shunting, will in all probability, assist an eastbound train up to Woodley before making its way back to the mid-Cheshire town where it was based. ***G Coltas***

59A. Tiviot Dale, 27th December 1957. One of the three Standard Class 4MT 2-6-0's allocated to Trafford Park when new, No **76089** comes to a stand alongside the Up platform with the 1.30pm ex-Manchester Central to Chinley train. These engines were frequently to be seen on the Chester via Northwich services as well, and on this occasion, those via Stockport, Chinley and the Hope Valley line to Sheffield Midland. This service however only ventured between Manchester and Chinley, calling at all stations, Chorlton, Withington, Didsbury and Heaton Mersey prior to a four minute stop at Tiviot Dale. From then on it was Romiley, Marple, Strines, New Mills Central and Chinley, surprisingly missing out Buxworth in an overall journey time of 1 hour and nine minutes for the 20 miles. ***R Keeley***

60. Tiviot Dale. 1962. Another view of the station front at Tiviot Dale,with the forecourt once again virtually empty. Although a lengthy building, the accommodation was of a limited nature, many of the original functions having been dispensed with. At one time refreshment rooms occupied much of the left wing of the main building. In the centre were the booking offices with passageways either side. The Station Master's Office was immediately to the right of the gabled centre section. In May 1962, Mr. Ramsdale was the Station Master and much of the spare accommodation had been claimed for departmental use by the Signal and Telegraph and Civil Engineering groups. Train services had continued to be rationalized, with ten each way on the Manchester Central to Derby or Sheffield via Romiley runs. Tiviot Dale still however, had its southbound through train to St.Pancras, the 12.05am ex Manchester Central, departing Stockport at 12.25am with arrival in London at 6.15am. On the Cheshire Lines, the number of through trains totalled five during the week, three from Warrington and two from Liverpool. On Sundays it was necessary to travel via Manchester Central although the weekend service was only marginally worse than that during the week. The architect of the station, a Mr Mangnall of Manchester, was undoubtdly not thinking of photographers when he finalized the position of the main buildings. Views of the entire elevation were difficult to achieve and although there were many attempts at photographing the pile, few achieved complete success. This view from the Stockport Libraries collection shows the elegant lines of the collonade, the gabled centre and end sections complementing the many arched arcading. According to an article in the *"Builder", after the 7th September 1866, the new station of the Stockport and Timperley Railway was progressing, preparation of the site requiring the removal of some 100,000 cubic yards of earth and rock.*Masonry, brick and excavation work was carried out by a local man, a Mr Forrester, with assistance from a Mr D. Cochrane of Manchester. The building is reported to have cost approximately £8000. *Stockport Libraries.*

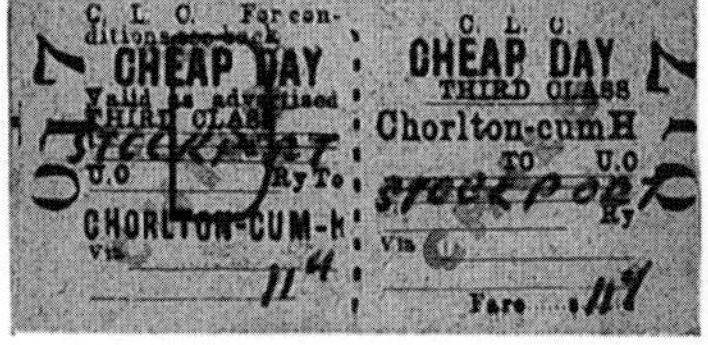

61. Tiviot Dale, 30th December 1966. No sign of any trains, but the perpetual veil of smoke which engulfed the tunnel end is very much in evidence. To the left of the picture a short bay platform, one of two on the Down side that had not seen passenger usage for many years. Certainly until the late 1940's an overall roof had offered protection against the elements. *W.J. Skillern.*

62. Tiviot Dale, May 1961. The Jubilee's were frequent visitors to the line during the 1950's although much of their work was of a very moribund nature. A fairly lightweight turn of duty saw No.**45607** *Fiji*, at the time a Millhouses (Sheffield) engine, with the 11.40am (SO) Manchester Central to Chinley train. This leisurely journey of twenty miles took all of 64 minutes, interrupted only by stops at Chorlton, Withington, Didsbury, Heaton Mersey, Tiviot Dale, Romiley, Marple, Strines and New Mills Central. Connection at Chinley was made with the 12.25pm ex Manchester Central to St.Pancras express, which had run non-stop via Cheadle Heath. After a twenty or so minute break *Fiji* and train would form the 1.05pm service to Sheffield via the Hope Valley. *H. Challoner.*

63. Tiviot Dale, May 1961. Saturdays at Tiviot Dale were no less interesting as far as freight traffic was concerned and the industrial area adjacent to Portwood provided a fitting backdrop to this short goods working, possibly the 11.50am from Dewsnap to Heaton Mersey Sidings. The locomotive, British Railways Standard Class 4 2-6-0 No.**76087** was one of a group of three completed at Horwich in May 1957. Initially allocated to Heaton Mersey, she soon set off for pastures new with a fifteen month stint at Saltley. A second spell at Heaton Mersey from January 1959 resulted in a three year stay which included such duties as that seen here. Apart from the concrete lamp standards, Tiviot Dale had resisted change for decades, and was to retain its individuality until its wasteful closure on New Years Day in 1967. *H. Challoner.*

64. Tiviot Dale, c.1962. This view looking west from the former Down bay platform shows a station which, with the exception of the concrete lamp standards had remained virtually unchanged for something like half a century. Today, all that remains, separated by rubble and assorted vegetation. The photographer would be several feet above the eastbound carriageway of the M60. ***G H Platt***

65. Tiviot Dale, 29th June 1968. The railway, now having become freight only, no longer needs the home comforts of the passenger station that served the town for just over a century. Eighteen months after the withdrawal of passenger services, the demolition men have moved in, reducing the building to a skeletal form prior the final act of complete removal. Contractors for the demolition work were Messrs Dawson and Morris of Skipton. This view is taken from a wall of the Christadelphian Church, opposite the station across Lancashire Hill. *Stockport Libraries.*

66. Tiviot Dale.16th December 1962. The cobbled approach from the east end of Tiviot Dale which, from this angle, belied its location high above the River Mersey, some fifty or so feet below the cast iron railings to the left. All the advantages of the location i.e. east of access, close proximity to the town centre, adjacent bus stops, adequate car parking etc; counted for little following Dr Beeching's pronouncements.
Stockport Libraries.

67. Tiviot Dale, 22nd July 1947. This view of the station from King Street bridge graphically illustrates the ledge on which Tiviot Dale was located. At this point, the River Mersey was formed by the confluence of the Goyt and Tame. The photograph also shows a little known feature of the station, an overall roof covering the bay platforms at the east end on the Down side. By this time its use was really no more than for carriage storage but in the earlier years could well have been used for the frequent train service between Stockport and Woodley. The 1885 timetable showed a weekday service of twenty two east bound trains and twenty one west bound for the 7 and 6 minute journey respectively. The timetable is reproduced below.
W.J. Skillern.

STOCKPORT and WOODLEY.—Cheshire Lines.

Tiviot Dale Sta.,	mrn	mrn	mrn	mrn	mrn	mrn	mrn	mrn	mrn	mrn	aft	aft	aft	aft	aft	aft	aft	aft	aft	aft	aft	aft	
Stockport ....dep	6 0	6 45	7 25	8 30	8 58	9 42	9 48	1045	11 4	1120	1220	1242	1 30	2 31	2 53	4 42	5 20	6 25	7 30	8 45	9 5	1010	....
Woodley 309, 314 arr	6 6	6 51	7 32	8 36	9 5	9 49	9 55	1052	1111	1127	1227	1249	1 37	2 38	3 0	4 49	5 27	6 32	7 37	8 52	9 12	1017	....
Woodley ......dep	7 47	7 55	9 43	1010	1032	11 3	1155	1 12	2 0	2 22	3 5	3 18	5 27	5 52	7 3	8 7	8 30	9 5	10 5	1025	1118	....	....
Stockport (T.D.) arr	7 54	8 2	9 50	1017	1034	1110	12 2	1 19	2 7	2 29	3 12	3 25	5 34	6 0	7 10	8 14	8 37	9 12	1012	1032	1125	....	....

[Sundays.

Tiviot Dale Sta.,	mrn	mrn	mrn	mrn	aft	aft	aft	aft	aft	
Stockport ....dep	8 50	9 22	9 42	1010	2 0	4 53	6 14	8 20	9 0	....
Woodley 309, 314 arr	8 57	9 29	9 48	1017	2 7	5 0	6 21	8 27	9 7	....
Woodley ......dep	9 8	9 43	1017	1028	3 48	7 22	8 17	9 57	10 5	....
Stockport (T.D.) arr	9 15	9 50	1024	1035	3 55	7 29	8 24	10 4	1012	....

68. Tiviot Dale. 29th August 1963. One of Trafford Park's (9E) longest serving engines, Fairburn 2-6-4T No.**42675**, prepares to leave Tiviot Dale with the 2.07pm stopping train from Manchester Central (dep.1.45pm) to Chinley via Romiley and Marple. This train formed the 2.55pm service from Chinley to Sheffield Midland - via the Hope Valley. It also provided a connection at Chinley with the 2.25pm Manchester Central to London St.Pancras express, which during the week ran as "The Palatine", by now the only titled train covering the former Midland line to Manchester. *G. Coltas.*

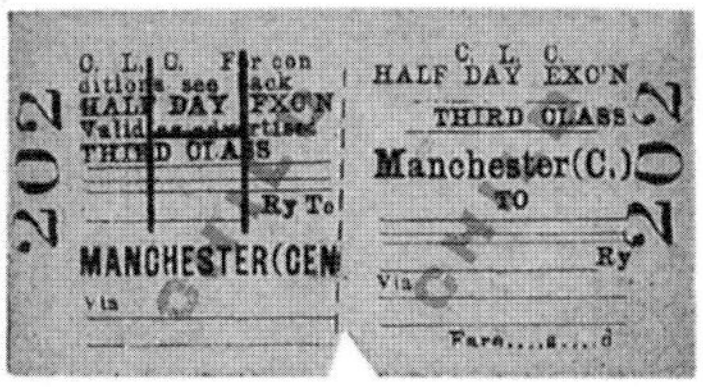

69. Tiviot Dale. 16th June 1962. One of the few photographs which show the platform elevation of the main station building. Unfortunately something has got in the way, on this occasion a Stanier 2-cylinder 2-6-4T No.**42598**, a Speke Junction (8C) engine ready to depart with a stopping train for Liverpool via the CLC. Until its closure in 1961, Brunswick shed had traditionally provided engines for the Stockport service. Numbers of the locomotives were dispersed between Speke Junction (Garston) and Heaton Mersey, continuing in use on the same services. This particular engine had a little over twelve months to go before being withdrawn towards the end of 1963. It had been one of a batch built in 1936 at the North British works in Glasgow. *A. Steele.*

70. Tiviot Dale. 21st March 1964. Visits by Liverpool based engines were to remain commonplace throughout 1964, last year of operation for the passenger services over the former Cheshire Lines system. This view shows a grimy engine in a rather all to familiar scene of neglect, perhaps compounded by the fact that winter has not yet thrown off its veil. One of lot 3536 built at Brighton Works in 1950, No.**42078** was a Fairburn development of the Stanier design, spending its earlier years on the Southern Region. In 1958, it was transferred to North Wales, where it was based at Croes Newydd and Bangor for short spells. Liverpool and the North West provided its running grounds for the last six years of its life, ultimately at Bank Hall, from where it was withdrawn in 1966, only sixteen years of age. *G. Coltas.*

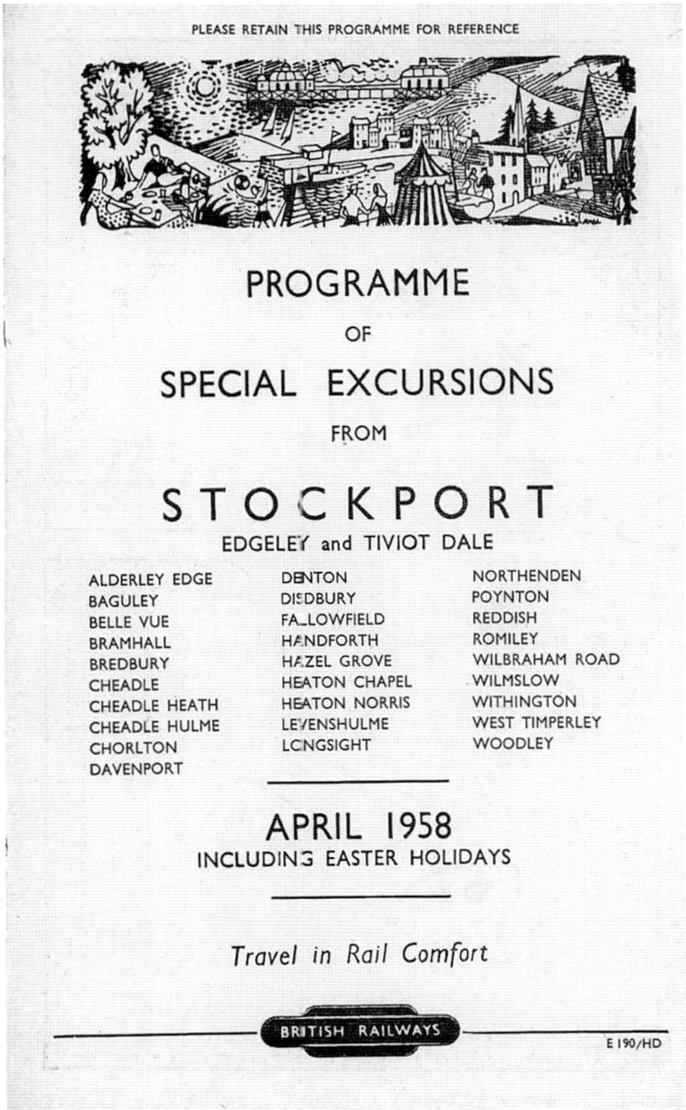

PLEASE RETAIN THIS PROGRAMME FOR REFERENCE

PROGRAMME

OF

SPECIAL EXCURSIONS

FROM

STOCKPORT

EDGELEY and TIVIOT DALE

ALDERLEY EDGE
BAGULEY
BELLE VUE
BRAMHALL
BREDBURY
CHEADLE
CHEADLE HEATH
CHEADLE HULME
CHORLTON
DAVENPORT
DENTON
DIDSBURY
FALLOWFIELD
HANDFORTH
HAZEL GROVE
HEATON CHAPEL
HEATON NORRIS
LEVENSHULME
LONGSIGHT
NORTHENDEN
POYNTON
REDDISH
ROMILEY
WILBRAHAM ROAD
WILMSLOW
WITHINGTON
WEST TIMPERLEY
WOODLEY

APRIL 1958

INCLUDING EASTER HOLIDAYS

Travel in Rail Comfort

BRITISH RAILWAYS

E 190/HD

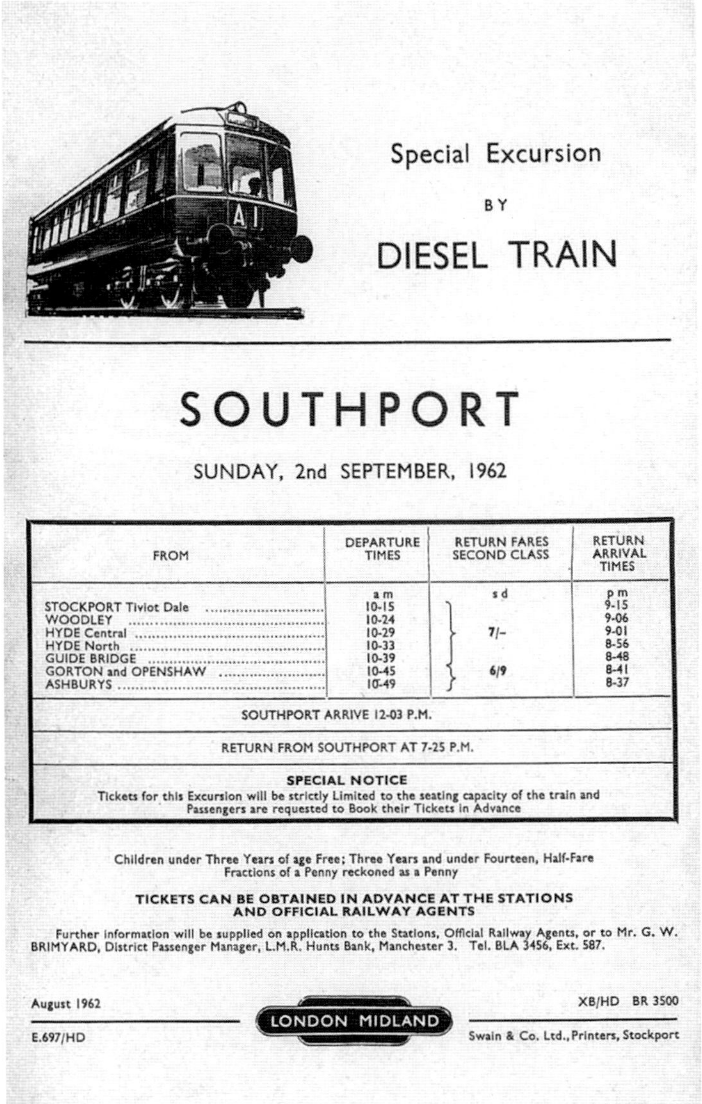

Special Excursion

BY

DIESEL TRAIN

SOUTHPORT

SUNDAY, 2nd SEPTEMBER, 1962

FROM	DEPARTURE TIMES	RETURN FARES SECOND CLASS	RETURN ARRIVAL TIMES
	a m	s d	p m
STOCKPORT Tiviot Dale	10-15	7/-	9-15
WOODLEY	10-24		9-06
HYDE Central	10-29		9-01
HYDE North	10-33		8-56
GUIDE BRIDGE	10-39	6/9	8-48
GORTON and OPENSHAW	10-45		8-41
ASHBURYS	10-49		8-37

SOUTHPORT ARRIVE 12-03 P.M.

RETURN FROM SOUTHPORT AT 7-25 P.M.

SPECIAL NOTICE

Tickets for this Excursion will be strictly Limited to the seating capacity of the train and Passengers are requested to Book their Tickets in Advance

Children under Three Years of age Free; Three Years and under Fourteen, Half-Fare
Fractions of a Penny reckoned as a Penny

TICKETS CAN BE OBTAINED IN ADVANCE AT THE STATIONS AND OFFICIAL RAILWAY AGENTS

Further information will be supplied on application to the Stations, Official Railway Agents, or to Mr. G. W. BRIMYARD, District Passenger Manager, L.M.R. Hunts Bank, Manchester 3. Tel. BLA 3456, Ext. 587.

August 1962

XB/HD BR 3500

LONDON MIDLAND

E.697/HD

Swain & Co. Ltd., Printers, Stockport

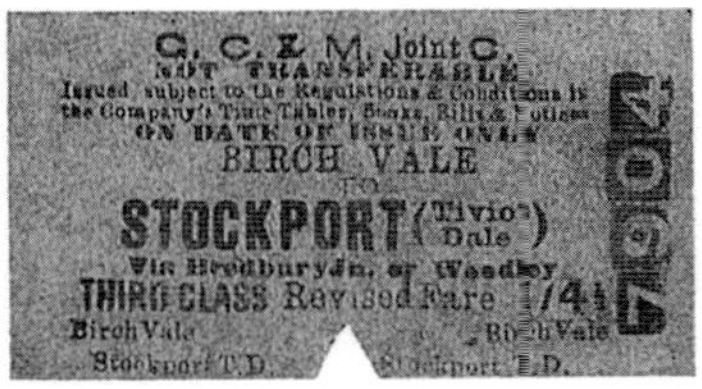

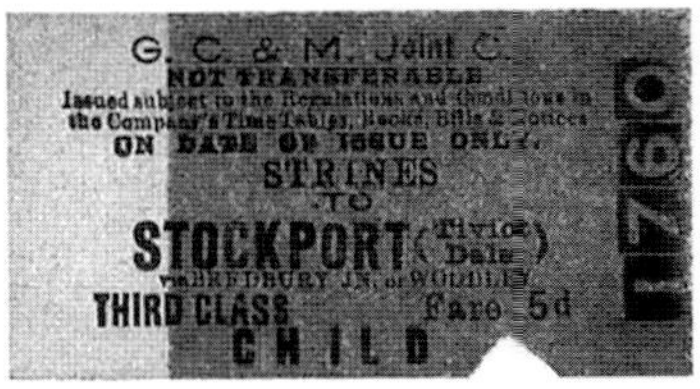

71. Tiviot Dale. 12th March 1962. It will have been something like three hours since this engine left Dee Marsh with its train of empty coal wagons. First stage of the journey will end at Godley where the locomotive, No.**63794**, Class O4/7, of Northwich shed will be turned before working back to her home depot. Although not exactly qualifying for the statement that this traffic was a constant stream, the trains were numerous. Barnsley Main, Sheffield Bernard Road, Wharncliffe, Warsop, Newstead, Kirk Sandall, Wombwell, Wath, were but eight of the many destinations. Much of the westbound coal traffic from Yorkshire was marshalled in the triangle at Halewood (Liverpool) prior to being taken to Herculaneum Dock for shipping to Ireland. *G. Coltas.*

72. Tiviot Dale, 21st March 1960. As mentioned previously, Jubilee Class locomotives were frequent visitors to the line at this time. One in particular, No.**45664** *Nelson* was more than familiar to trainspotters, particularly on the Liverpool to Nottingham service. However, at the time of the photograph, she was a Millhouses (41C) engine and more than likely on one of the Sheffield bound trains, the timetables of the relevant service being reproduced below, Bradshaw 7:3:60 to 3:4:60.

courtesy A. Barnard.

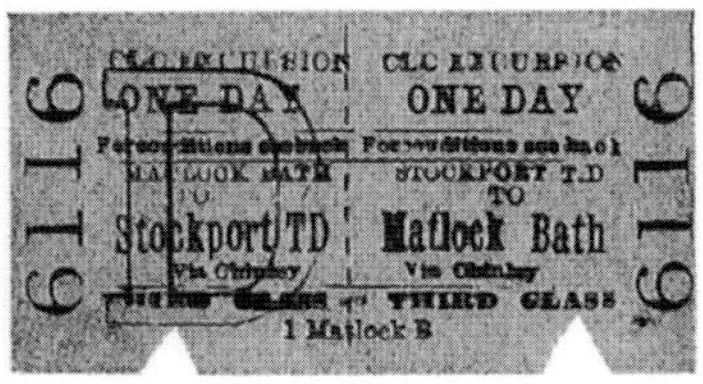

Table 212 MANCHESTER, CHINLEY, HATHERSAGE and SHEFFIELD

Miles		Week Days																	Sundays												
		am	am	am		p.m		pm	pm E C	pm E T	pm S T	pm B		pm	pm		p.m E T	pm S T		a.m		am T		pm		pm D		p.m H	pm		pm	
	Manchester (Cen.) dep	..	6 5	8 55	—	1225	..	2 25	3 25	4 5	4 5	4 25		5 55	7 45	..	9 30	9 30		8 55	..	1016	..	1 55	..	..	..	5 4	5 55	..	7 10	..
	Stockport (T.D.) "	..	6 34	8 42	—	12 9	..	2 1	3K48	4 31	4 35	4 52	..	6F35	8T 0	..	9 56	9 56	..	[illegible]	..	[illegible]	..	1 55	..	..	..	5 21	5 29	..	..	..
20	Chinley "	7 0	7 40	9 47	..	1 5	..	3 10	4 25	5 8	5 12	5 40	..	[illegible]	8 30	..	1045	10 45		[illegible]	..	1112	..	3 17	..	..	..	6 12	6 45	..	8 17	..
25½	Edale	7 19	7 53	9 56	..	1 19	..	3 24	4 38	5 21	5 25	5 53	—	6 46	8 43	..	..	10 59	..	[illegible]	..	1129	..	3 29	..	..	..	6 26	6 59	..	8 29	..
30¼	Hope A	7 21	8 2	10 4	..	1 27	..	3 32	4 40	5 29	5 33	6 1	..	6 54	8 51	—	..	11 7	..	10 7	..	1134	..	3 37	..	6 35	..	6 34	7 7	..	9 37	..
32½	Bamford	7 25	8 6	10 8	—	1 31	..	3 36	4 50	5 32	5 37	6 5	..	6 58	8 55	..	..	11 11	..	1011		1138	..	3 41	..	6 38	..	6 38	7 12	..	9 41	..
34½	Hathersage	7 29	8 10	10 12	—	1 36	..	3 41	4 54	5 37	5 41	6 9	..	7 2	8 59	..	..	11 16	..	1016	..	1143	..	3 46	..	6 43	..	6 43	7 18	..	9 46	..
36½	Grindleford	7 34	8 15	10 17	..	1 41	..	3 46	4 59	5 42	5 46	6 14	..	7 7	9 4	..	..	11 21	..	1021	..	1148	..	3 51	..	6 48	..	6 48	7 23	..	9 51	..
41½	Dore and Totley	7 46	8 25	10 27	..	1 51	..	3 56	5 9	5 52	5 56	6 24	..	7 17	9 14	..	..	..		[illegible]	..	..	..	4 1	..	6 58	..	6 58	7 33	..	9 1	..
49¾	247 Chesterfield (Mid.) arr	8 12	8 44	..	..	..	..	4 39	5 57	6 20	6 20			8 9	9 52	..	..	..						4 26						..		
71	234 Mansfield (Town) 234a "	..	..	..	..	..	..					..	..				..	..														
42	Beauchief	..	..	..	—	..	..	3 59	5 12	5 55	5 59		..	7 20		..	..	..	..		..	..	..	..	..	..	..			..	..	..
42¾	Millhouses & Ecclesall	7 49	..	10 31	—	1 55	..	4 2	5 15	5 58	6 2	6 28	..	7 23	9 18	..	..	..		1026	..	..	..	4 5	..	7 3	..	7 2	7 37	..	9 5	..
44	Heeley	7 53	8 31	10 35	—	1 59	..	4 6	5 19	6 2	6 6	6 32	..	7 27	9 22	..	..	11 34		[illegible]	..	..	..	4 9	..	7 6	..	7 6	7 41	..	9 9	..
45¾	Sheffield (Mid.) arr	7 57	8 35	10 39	..	2 3	..	4 10	5 23	6 6	6 10	6 36	..	7 31	9 26	..	1126	11 33		[illegible]	..	[illegible]	..	4 13	..	7 10	..	7 10	7 45	..	9 13	..
61	247 Rotherham (Mas.) arr	8E23	8 51	11 13	..	2 40	..	4 24	5 41	6 42	6 42	7 0		8 32	9 45	..	1 41	1 41				1233		6 27	..	7 47		7 47			9 33	..

Miles		Week Days																		Sundays												
		am		am	am T		a.m		p.m S	p.m		pm B	pm		pm T	pm E		pm	pm		a.m		am	am K		am T		pm T	p.m H		pm T	pm	
	247 Rotherham (Mas.) dep	4 49	..	..	7 15	..	9 15	..	12 0	12E6		1 56	3 33	..	5J10	5 31	..	6 27	8 49		4 33		8 24	..	..		..	1 20		..	5 50	6 29	..
—	Sheffield (Mid.) dep	6 5	..	7 2	7 50	..	9 38	..	1235	1250		2 22	4 30	..	5 30	6 0	..	6 55	9 24		8 30	..	9 0	9 30	..	10 0	..	2 0		..	6 10	7 0	..
1¼	Heeley	6 10	..	7 8	7 55	..	9 43	..	1240	1255	..	2 27	4 35	..	5 36	6 5	..	7 0	9 29		[illegible]	..	9 6	9 36	..	[illegible]	..	2 7	..	..	6 16		..
2¾	Millhouses & Ecclesall	..	..	7 14	8 0	..	9 48	..		1 0	..	2 32	4 40	..	5 42	6 10	..	7 5			8 42	..	9 12			1012	..	2 13		..	6 22		..
3½	Beauchief	..	..	..	8 3	..	9 51	..		1 3		2 35	4 43	..													..						..
—	234a Mansfield (Town) 234 dep	..	..			..	8B56	..						..			..										..						..
—	247 Chesterfield (Mid.) "	..	..	7 0	7 24	..	9 9	..				1 32	4 1	..	5 15	5 50	..		8 30		7 55				..		..						..
4¾	Dore and Totley	6 19	..	7 21	8 6	..	9 54	..	1248	1 6	..	2 38	4 46	..	5 48	6 17	..	7 10	9 37		[illegible]		9 19			1018	..	2 20	..	..	[illegible]		..
9½	Grindleford	6 30	..	7 32	8 16	..	10 4	..	1258	1 16	..	2 48	4 56	..	5 59	6 25	..	7 20	[illegible]		9 2	..	9 30	9 54	..	[illegible]	..	2 31	..	..	6 39	7 21	..
11¼	Hathersage	6 35	..	7 37	8 21	—	10 9	..	1 3	1 21	..	2 53	5 1	..	6 4	6 30	..	7 25	9 52		9 8	..	9 37	9 59	..	[illegible]	..	2 36	..	..	6 44	7 26	..
13	Bamford	6 40	..	7 42	8 25	..	10 13	..	1 7	1 25	..	2 57	5 5	..	6 9	6 34	..	7 29	9 56		9 13	..	[illegible]	10 4	..	[illegible]	..	2 41	..	..	6 49	7 31	..
14¾	Hope A	6 47	..	7 48	8 31	..	10 19	..	1 13	1 31	..	[illegible]	5 11	..	6 15	6 40	..	7 37	10 2		..	..	9 48	1010	..	1045	..	2 47	5 25	..	6 54	7 37	..
19¾	Edale	6 59	..	8 1	8 43	..	10 31	..	1 25	1 43	..	3 17	5 23	..	6 29	6 52	..	7 47	1014		[illegible]	..	10 1	..	..	1058	..	[illegible]	5 37	..	[illegible]	7 50	..
25¾	Chinley arr	7 11	..	8 13	8 55	..	10 43	..	1 37	1 55	..	3 27	5 35	..	6 40	7 4	..	7 59	1025		[illegible]	..	1013	..	..	1110	..	[illegible]	5 49	..	7 21	8 2	..
37¾	Stockport (T.D.) "	7L47	..	8L42	9Z34	..	..	..	..	3§27	..	..	6 36	..	7 17	..	..	8 48	1116		..	..	..	..	..	[illegible]	..	3 51	..	..	[illegible]	[illegible]	..
45¾	Manchester (Cen.) "	8 5	..	8 59	9 24	..	11 23	..	..	2 31	..	..	7 5	..	7 41	..	..	8 40	11 5		..	..	..	..	..	1211	..	4 15	..	..	8 17	8 45	..

A For Castleton (2 miles) and Bradwell
B Saturdays only. Commencing 16th April
B Via Pye Bridge
C Through Carriages Stockport (T.D.) to Sheffield
D Runs 17th and 24th April
E or E Except Saturdays
F Except Saturdays. Cheadle Heath (Stockport). Passengers can dep Tiviot Dale 4 51 pm (daily)
H Commences 1st May
H Cheadle Heath (Stockport)
J Dep 4 29 pm on Saturdays
K Commences 17th April
K Passengers can dep Cheadle Heath 8 40 pm
L Cheadle Heath (Stockport). Passengers can arr Tiviot Dale 8 47 am
S or S Saturdays only
T Through Carriages between Manchester and Sheffield
T Cheadle Heath (Stockport). Passengers can dep Tiviot Dale 7 16 pm
§ Arr 3 25 pm on Saturdays
Z Passengers can arr Cheadle Heath (Stockport) 9 54 am

73. Tiviot Dale. 1st June 1957. As mentioned previously, the former Midland 3-cylinder Compounds were frequent visitors to the line, particularly on the Manchester Central to Sheffield/Derby services. Although No.**40927** makes a truly evocative departure from the Up platform, she had only a few days of life remaining, her freedom ending at the place where she had been conceived, Derby Works. Perhaps fittingly, she is seen leaving with a stopping train, the 2.50pm Manchester Central to Derby, calling at quite literally - Strines and Derby, Nottingham Road excepted - every station on the route. In an era when one could travel for the pleasure of travelling, the three and a half hours this train took for the 61½ mile journey must have been filled with delight as it passed leisurely through the scenery on offer in the Peak District. On the other hand, those actually wanting to get to Derby could leave Manchester over one hour later (on the 4.00pm to St.Pancras) and arrive at 5.38pm, fifty minutes earlier.

courtesy N.E. Preedy.

74. Tiviot Dale, 5th October 1959. Probably the most familiar engine in the 1950's on freight workings was the Class O4. 2-8-0 locomotive of Great Central origin. This particular example was classified O4/3 to a design introduced in 1917. Her association with Gorton was long, retaining it until withdrawal in August 1962. Judging by the ladders leaning against the signal box, the structure is to receive the customary coat of paint, the results of which can be seen below. *R. Keeley.*

75. Tiviot Dale. 1962. Standing sentinel like at the east end of the station, this all timber CLC box acted as guardian to the thousands of passing trains from the early years of the 20th Century until its closure in 1967. The name on the end of the box displays simply "Stockport", the powers that be obviously not concerned about that other line through Edgeley. Until closure of West signal box, this had carried the name Stockport Tiviot Dale East. This forty four lever signal box, once the passenger services had ended, was quickly made redundant. Closure came on 1st July 1967, its control arrangements being shared between Georges Road and Brinnington Junction signal boxes. Consequently train banking commenced at Georges Road instead of Tiviot Dale with the platform loops being taken out of use. *G.H. Platt.*

76. Tiviot Dale. 1962. The view east from the end of the Up platform with carriages stored in the bay. The signals, no doubt originating from the CLC's own workshops at Warrington, remained in use until abolition of Tiviot Dale as a signalling block post in 1967. The signal arms, from the left to right, controlled movements from the bay platform, Up platform and Up through line respectively. They were operated by levers 38, 42 and 39 in the signal box. The gantry in the distance replaced ageing Cheshire Lines equipment in 1959. In this view, a signal (lever 3) is off, clearing access into the Down platform. *G.H. Platt.*

77. Tiviot Dale, 1952. Banking duties at Tiviot Dale were carried out by a great variety of motive power. The older generation will fondly recall the former Great Central tank engines that performed the task, the N5 0-6-2 tank being mentioned specifically. Tucked away in the Up bay platform is No.**69359**, standing ready to assist when required. This veteran was of a design introduced in 1891 and spent her final years at Heaton Mersey. *courtesy A. Barnard.*

78.Tiviot Dale. 24th April 1962. Not quite a stranger to the territory but certainly a change from the bulky outline of the former Great Central locomotives traditionally associated with the east/west freight movements. On this occasion, former LNER B.1. Class 4-6-0 No.**61077**, a Northwich (9G) based engine, takes the through line with a class 6 - partially braked - freight train on its way to Walton (Warrington). Having left Mottram at 9.27am, the B.1. will journey via Skelton Junction and the Lymm line prior to arriving at its destination. Like numerous other members of the class, this locomotive had been transferred north following the run down of the former Great Central main line services. It had seen duty at Neasden (14D) and Leicester (15E) before coming to mid-Cheshire. *G. Coltas.*

79. Tiviot Dale. 8th September 1965. It was at this point in Tiviot Dale where the River Tame flowed into the Mersey, although the substantial bridgework required to span the two sides is unfortunately not reflected in this view from King Street bridge. Needless to say a quite large structure emerged, having a span in excess of 100 feet. The outer members comprised two deep wrought iron box girders, with equally deep lattice girders of the same material in between. Unfortunately the somewhat flimsy looking handrails spoiled the overall effect and from the top looked just like any other bridge. *G. Coltas.*

80. Tiviot Dale. 25th July 1963. Another wet summers' day in Stockport as ex LNER B.1 Class 4-6-0 No.**61109** gets to grips with the 3.53pm train for Chinley and Sheffield. This weekdays only service left Manchester Central at 3.30pm and called at all stations - Strines excepted - on its way to Sheffield via the Hope Valley. Once again we have an engine that was far removed from the location at which it had seen its better days. For many years Stratford (30A) had been its home depot, no doubt hauling many an express to and from East Anglia. Latterly it had been transferred to the Sheffield area and was at the time of the photograph based at Sheffield (Darnall). *G. Coltas.*

81. Tiviot Dale. 30th December 1966. Having spent much of the previous decade based in the East Midlands it is unlikely that this Stanier Class 5 would suffer much of a culture shock on being transferred to Trafford Park. This view shows No.**44851** taking the Down through line with a rather lightweight local freight, initially to Heaton Mersey Sidings. Although freight was to continue for a good number of years, the station as a passenger operation had just two days remaining. The three dwarf signals to the right of the locomotive were for allowing movements against the normal flow of traffic. Some time during the 1950's they had replaced a 3 doll (signal) balanced bracket located approximately mid way between the second and third discs, in fact it was almost a mirror image of the bracket on the opposite platform. *G. Coltas.*

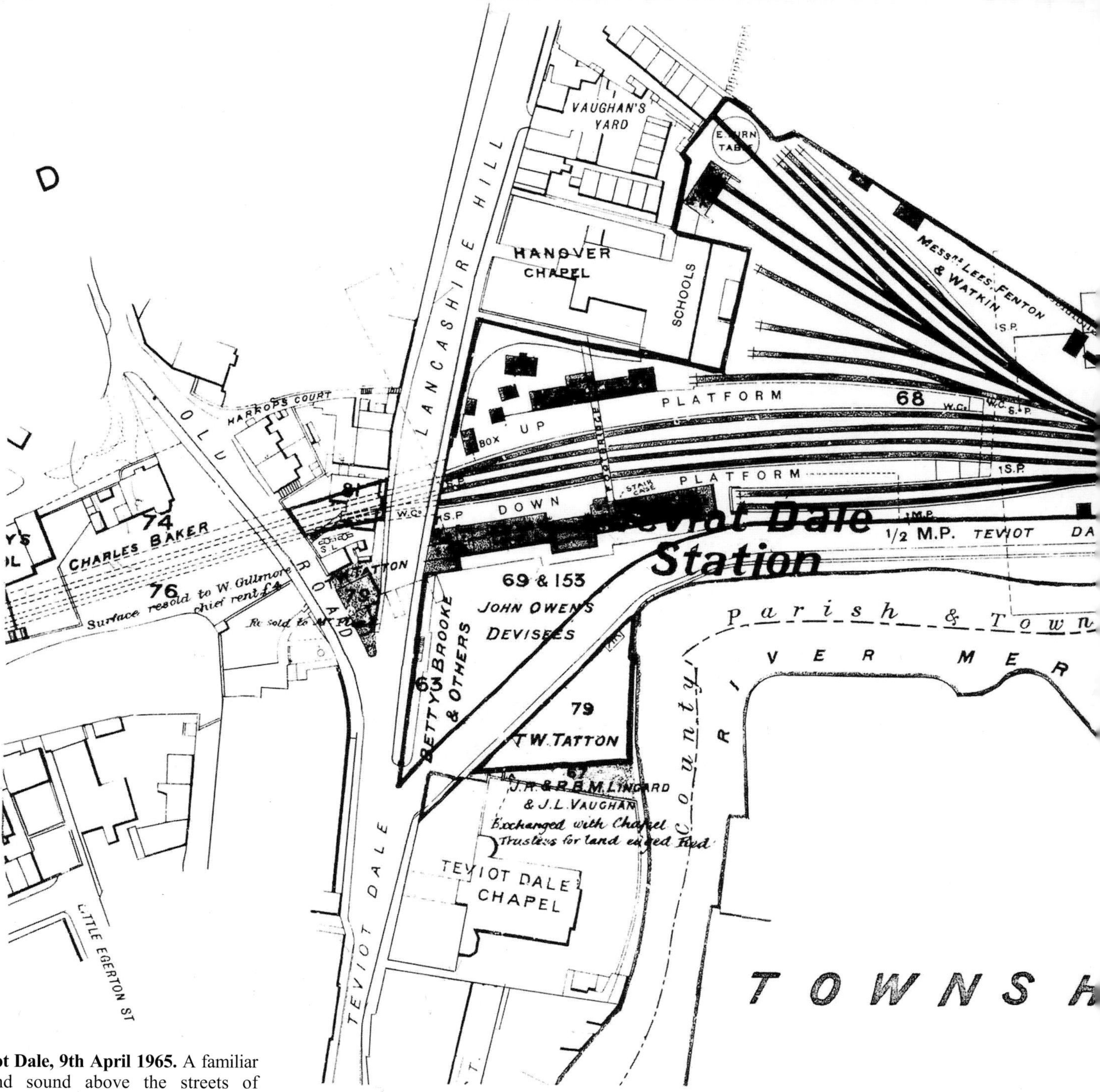

82. Tiviot Dale, 9th April 1965. A familiar sight and sound above the streets of Portwood was the passenger train working hard on the initial climb from the station. This working is the 1.45pm from Manchester Central to Chinley which took in all stations via Romiley and Marple, reaching its destination at 2.40pm to provide a connection with the southbound 'Palatine'. The structures seen here were by now over one hundred years old although those to the left had been strengthened or reconstructed as necessary. The first carriage is passing over bridge No 84, Portwood Place, whereas the engine, No **75042** is on the concrete span over Pool Lane. The Ford 'Popular' is travelling along Avenue Street at a point now occupied by the eastbound carriageway of the M60. These openings were the westerly end of Portwood Viaduct, demolished in the 1990's as far as Mottram Street (now Tiviot Way) to enable redevelopment of the land, and consequently breaking the integrity of the route. ***G. Coltas***

13 Miles

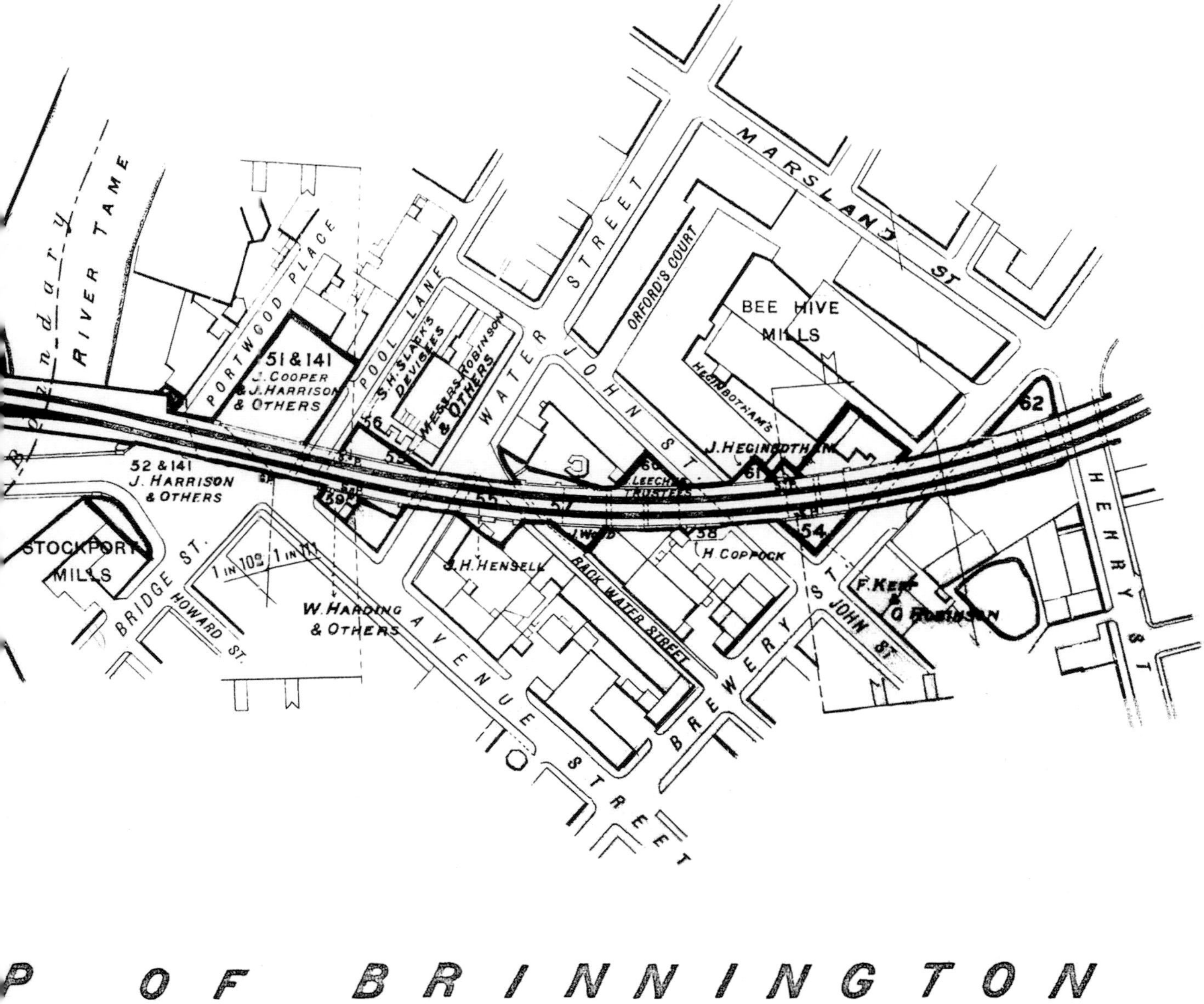

83. Tiviot Dale. 1st June 1957. A combination of gradients, curvature and condition of structures served to restrict the speed of trains. The section east of the station, although known as Portwood Viaduct, was actually a series of independent bridge, all numbered in sequence in a west to east direction. It was here that a twenty miles per hour speed restriction was placed in both directions, although Down, or westbound trains usually threw caution to the wind as they passed through the station in the knowledge that they had a short but severe series of adverse gradients to reach Georges Road. Here we see trains cautiously passing each other, the passenger working having just left the station, (see plate 73) commences its ascent to Brinnington Junction whilst the goods, in the hands of Class O.4 2-8-0 No.**63603**, announces its presence by making smoke in an area of the town which, at the time, would not be out of keeping.

courtesy N. Preedy.

83B. (Above) Tiviot Dale, c.1964. Another of Heaton Mersey's allocation, this time in the form of Fairburn Class 2-6-4T No **42134**, stands alongside the Up platform with the 4.56 ex-Liverpool Central train, due in Tiviot Dale at 6.34pm after a $37\frac{1}{4}$ mile journey which started with a non-stop run to Hunts Cross. It was then all stations via Warrington and Glazebrook where connection was made with the 5.5pm ex-Wigan Central to Manchester Central. ***G Coltas***

83C. (Centre) Tiviot Dale, July 1949. The motive power of yesteryear that plied the Stockport to Liverpool route in the immediate post war period was in the form of Great Central Class D9 4-4-0 No **2311**, still carrying its LNER livery and number. The train is on the Down Through line prior to repositioning in the platform road. The overall journey was marginally quicker than by later services. This useful link with Warrington and Liverpool was withdrawn on 30th November 1964. ***F Dean***

83D. (Right-) 1st January 1967. Not a happy new year for Tiviot Dale and its users as this was the last day of operation before withdrawal of services from the 2nd. Passengers taking advantage of the restaurant car (M333) are caught by the wintry sunshine as the 8.45 ex-London St Pancras to Manchester Central train comes alongside the platform some 4hrs and 24mins after leaving the capital. Its the journey that matters one might say as numerous of these passengers would have been sitting to lunch whilst travelling through the Peak District. Within a minute of departure from Tiviot Dale, passengers for Didsbury would also be getting ready to alight, for the last time! ***W J Skillern***

83E (Above) Tiviot Dale, 22nd May 1962 (Tuesday). Reduced to almost a crawl by the imposition of a 20mph speed limit, the 10.35am ex-Derby to Manchester Central approaches Tiviot Dale station behind Sulzer engined type 4 (later Class 45) 1Co-Co1No **D122** - at the time barely six months old - and Derby based Class 5 4-6-0 No **44825.** This was an all stations (Derby Nottingham Road excepted) to Tiviot Dale train. On the last 'lap' into Manchester, only Didsbury and Chorlton by this time remained open.However, the overall journey time during weekdays was two minutes longer than the Saturday service. ***G Coltas***

83F (Centre) Tiviot Dale, c.1966. A view from the window of the leading carriage on this Derby-Manchester stopping train shows the approach to Tiviot Dale in the final months of local passenger services on the route. Above the engine, a banner repeating signal has been installed to help overcome sighting difficulties of signals ahead. ***M S Cross***

83G (Below)Tiviot Dale, 29th April 1963. A cautious approach is demanded by the driver of Class 4F No **44289** (9F-Heaton Mersey) with his trainload of coal from Godley Junction to Partington coal basin. The train has the all-clear for the Through line from the gantry signal. ***G. Coltas***

Bridges played a more than significant part in carrying railways through the town. Whilst the monumental Stockport Viaduct took pride of place in carrying the Manchester & Birmingham from one side of the Mersey valley to the other, an equally formidable task faced the promoters of the Stockport, Timperley & Altrincham Junction Railway in trying to secure an east to west route to link up with the Stockport & Woodley line at Portwood. After three tunnels and a substantial rock cutting, progress east still required a series of eighteen bridges to link Tiviot Dale and Brinnington. Most were plain brick arches but the two shown here were small wrought iron types with decorative cast iron parapets. Numbered in sequence from Glazebrook, the first three views show bridge No 88 carrying the railway over Water Street. Everything in view nowadays has been razed to the ground.

84. Tiviot Dale. 21st July 1953. This view east shows the railway at track level from a point just before Water Street bridge (No.88). The distinctive cast iron parapet girders are seen to good effect although the decorative patterns were on the outer surfaces. To the left of the signal post is the spire of the now demolished St.Paul's Church which was located on the corner of Carrington Road/ Brinnington Road. *British Railways.*

85. Tiviot Dale. (Water Street). 21st July 1953. This street level scene illustrates well the 1950's urban decline evident in the Portwood area. The two small children playing in front of the terraced houses was a familiar feature in the tightly knit community. Few of these buildings survive today although in the distance, the outline of Kershaw's Tannery still forms part of the industrial skyline. The well known stepped profile of Stockport's street name signs can be seen adjacent to the bridge abutment. The name **Backwater Street** perhaps says it all! Opposite the dwellings was the Bee Hive, a public house which was originally one licensed to sell beer produced by Bells Hempshaw Brook Brewery. In 1949, Frederic Robinson Ltd., purchased Bell's and subsequently closed the Bee Hive. In 1960 it became head office of Parkside Precision Ltd. *British Railways.*

86. Tiviot Dale, 21st July 1953. The other side of Water Street bridge with Avenue Street at the far end. Through the opening to the left is another of Stockport's now forgotten street corner public houses. This was the Mechanics Arms, one time purveyors of beers from the Waterloo Brewery of Henry Smith & Sons in Hall Street. Nowadays we have a site devoid of the buildings and railway, replaced by, wait for it, the motorway.

British Railways

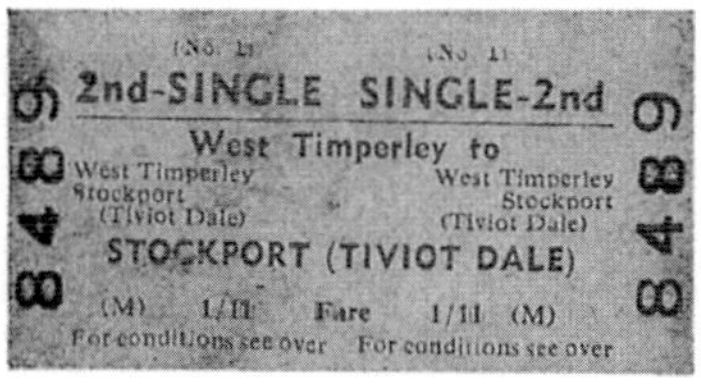

(Centre) 87. Portwood, 13th February 1963. A few yards further east, 'Crab' 2-6-0 No **42792** 9G-Gorton) crosses Marsland Street bridge with a Heaton Mersey to Dewsnap freight.This working was target 79 which left Heaton Mersey at 3.20am for a forty-five minute trip to the yard just east of Guide bridge. After disposing its wagons the engine departed light at 10.05 for godley Junction where it would pick up a westbound goods train to form the 11.40am to Portwood. The trip then continued with another short working, this time the 12.30pm portwood to Langwith Junction as far as Godley with a train comprising empty coal wagons. At the weekend it was engine and brake van only, with a return to Heaton mersey being arranged by taking the 1.30 freight as far as the Down sidings and then going on shed.

88.Portwood, 18th September, 1965. Again crossing Marsland Street but running wrong line is Class 9F No **92117**, a Speke Junction engine, with an east bound freight, To the right are buffer stops of the erstwhile Portwood coal drops, a small group of three sidings that could accommodate 84 wagons. These later became the site of the stone terminal, the infrastructure becoming the last of the railway formation being removed in 2003/4 to make way for a Tesco retail outlet.

Both photos; ***G Coltas***

89. Portwood. 21st July 1953. This excellent photograph was taken to record the location of bridge number 99, Marsland Street, although the wealth of detail in view provided a rare opportunity to review other structures and facilities in the Portwood area. The short lived Portwood station was situated in the area between the bridge and the warehouse. Access to the station was from Marsland Street (named Henry Street originally). This bridge however, marked the point at which the Stockport & Woodley Junction, and Stockport, Timperley and Altrincham Junction railways were linked to form the through route in 1865. Portwood station remained open for another ten years, closure taking place on 1st September 1875. To the left of the picture are the coal shoots mentioned previously whilst on the right is the large goods warehouse which unfortunately hides the even larger cotton shed that stood in the area bounded by George Street and Mottram Street (now Tiviot Way). In the distance it is just possible to establish the whereabouts of Brinnington Junction signal box and it was between there and the warehouse that three loop lines existed for westbound trains to set aside, mostly for attention to wagon brakes but occasionally to let other trains pass. These loops each had a capacity for engine, brake van and between 60 and 80 wagons respectively. The small group of sidings adjacent to the cotton shed also had room for 64 wagons. *British Railways.*

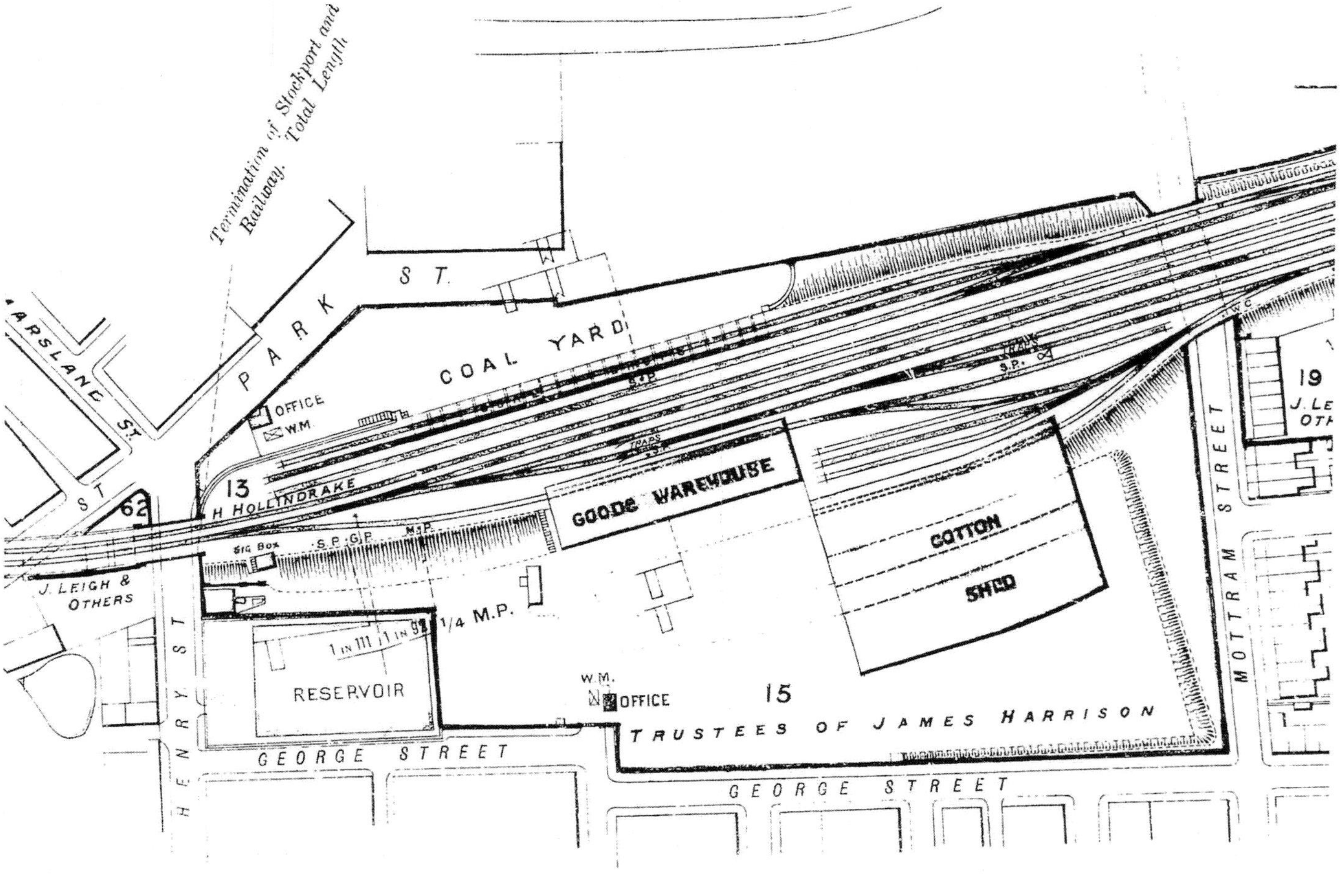

90. Portwood, 21st July 1953. This view from the south side of Marsland Street is of interest in that it shows the decorative cast iron parapets mentioned previously. The entrance to portwood Station was located to the right of the wing wall of the bridge.Once again, the main feature of this view is the row of houses in brewery Street amidst an almost overpowering background of industry.Whether name changing had become an obsession with the borough council is unclear but in addition to the change from Henry to marsland Street, the section seen here was formerly Park Street. The brewery referred to was Portwood brewery, established in 1796 (closed 1866) and located alongside the river behind buildings that were to become the base for Howards cycle and (later) motorcycle shop, now located in Newbridge Lane. Marsland Street is now the only through road passing beneath the formation of the erstwhile railway before Tiviot Way.
British Railways

(Left-centre) 91. Portwood, 21st July 1963. This time the north side of Marsland Street bridge with Great Portwood Street in the distance. The small shops, on a site now occupied by the Peel Centre retail development, highlight a scene gone forever, along with the communities which supported them, Swan Street, Queen Street, Emperor Street, Victory Street, to name but a few. The retaining wall to the left marked the boundary of the coal yard alongside Boundary Street. The bridge is now long gone and the whole area levelled and redeveloped.
British Railways

92. Portwood, 1st July 1963. Crossing Marsland Street bridge with the 3.30pm Manchester Central to Chinley/Sheffield stopping train is Class B1 4-6-0 No.**61313.** Although still carrying its 41A (Darnall) shed plate, the locomotive had recently been transferred to Canklow (41D). Leaving Tiviot Dale at 3.53pm this Saturdays excepted train missed out only Strines on its journey of 1 hour 51 minutes. *G. Coltas.*

93. Portwood. 13th February 1963. The photograph suggests plenty of action as this pair of 4F's set about their tasks, No.**44282** to the left performing banking duties whilst Heaton Mersey's No.**44501** organises the wagons on the coal drop sidings. At this point the climb towards Woodley begins in earnest with the 1 in 92 gradient steepening to 1 in 80 when passing Brinnington Junction. The three sets of lines to the left were loops numbered 1 - 3, and it is quite obvious that the third did not have a priority when it came to alignment and packing. *G. Coltas.*

(Centre) 94. Portwood, 26th August 1963. A daily feature at Portwood Sidings was this visit by a Heaton Mersey Class 4F 0-6-0 to carry out duties as the 'Portwood Shunt'. The trip, known to engine crews as Target 74, left heaton mersey shed at 7.50am, arriving portwood at 8'o'clock. It was covered by Turn 221, the men having come on duty at 7.35am. The engine shunted the portwood area until 2.40pm when the crew were relieved by the men of Turn 222. Work continued until 4.30pm at which point the engine was worked back into Tiviot dale to act as banker for trains to Woodley. At 9pm, this crew was relieved by the men of Turn 223 who had booked on at 8.15pm. Their hours of duty lasted until 4.25 the next morning. This activity was repeated for almost another three years before the goods warehouse closed in the Spring of 1966. The warehouse itself was demolished in February 1968, the site now occupied by a westbound access road on to the M60. ***G Coltas***

(Right-lower) 95. Brinnington Junction, c 1954. This all timber CLC pattern signal box controlled movements to and from the line to Reddish Junction, the junction itself seen here to the left of the train, a stopping train to Manchester Central. The working timetable for summer 1959 showed 3 westbound workings over the branch, all to Heaton Mersey, one being a light engine from Ancoats. Eastbound operations for some reason allowed for the passage of 4 workings, one of which was a daily Class E train to Hellifield. The line between Reddish Junction and Brinnington Junction, known sometimes as the Portwood Branch, was closed on 7th September 1965 and redundant materials were removed in 1966. The serviceable double junction at the Brinnington end was recovered and sent to Lenton (Nottingham) for storage. Brinnington Junction boc had a 40 lever frame, No 38 of which allowed operation of the 4 lever Portwood Sidings ground frame, which controlled access to the sidings of the yet to be built factory of T. Storey, manufacturers of the world famous 'Bailey' bridge. ***C H A Townley***

During 1931, the shorter of Brinnington's two tunnels was opened out. The Cheshire Lines Railway, under the supervision of Mr K.C. Marrion, A.M.I.C.E., arranged for the event to be photographed. The contractors were Messrs A. Monk & Co., Padgate, Warrington. Removal of earth and other materials from above the tunnel had taken place by 14th June (**plate 96**) in readiness for the demolition, which was to be carried out using explosives (**plate 97**). The last section was ready for demolition at 6' o'clock on the morning of Sunday 12th July. Three hours later, the rubble had been loaded into wagons (**plate 98**) and the Up line cleared (**plate 99**) to allow passage of the "Hikers" train, due to leave Tiviot Dale at 9.02am for Sheffield and the Hope Valley. *Author's collection.*

96.

97.

98.

99.

100. Reddish Junction. c.1957. The line from Brinnington Junction skirted the western side of what became Brinnington housing estate, the Reddish (North) to Bredbury section being joined immediately east of Reddish viaduct. The location was instantly recognizable to passengers travelling on a Manchester London Road bound train as the wheels clattered across the junction. Frozen points in winter occasionally caused "delays" and the author, as a schoolboy, remembers well the snowball "fights" with signalmen and traincrew whilst other railway staff cleared the line. Reddish Junction box closed on 18th November 1973 following the installation of colour-light signalling between Ashburys East Junction and Romiley Junction. *W.A. Brown.*

101. Brinnington, c.1965. The skies are blackened by the exhaust from another Stanier 8F 2-8-0 as it climbs towards Bredbury Junction with empty coal wagons. The spot between here and Lower Bredbury was completely cleared to allow construction of the motorway, even the partially obscured piece of ancient high ground locally known as the 'camels hump', a one-time favourite recreational area of local youngsters. The track bed has now been converted to a cycle route. ***John Fairclough***

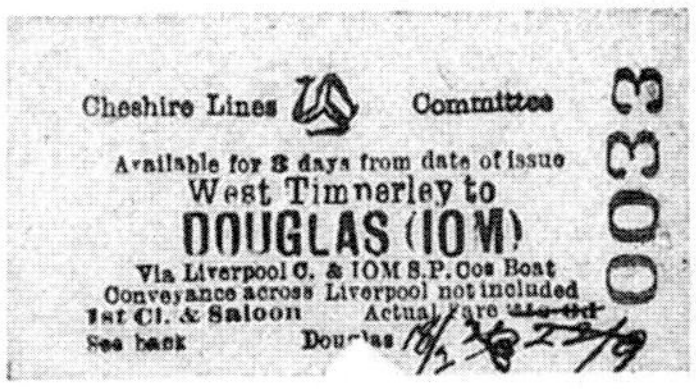

(Centre) 102. Brinnington, 1967. Viewed from above the entrance to Brinnington Tunnel, yet another Class 8F rolls steadily down the gradient towards Stockport, making sure it will be able to adhere to the 30 mph speed restriction around the curve at the other end of the tunnel. Today's view is one of forrestation, with the formation having been severed in the 1980's just beyond the rear of the train. By this time however the route had been singled and the coming of the then M63(now M60) sealed its fate. ***M S Stokes***

103. Brinnington, 1967. Stanier Class 5 4-6-0 No.**45046** approaches the eastern entrance to Brinnington (No.2) Tunnel, allowing the loaded coal wagons to produce the impetus for carrying the train down the 1 in 80 gradient. This 168 yard long tunnel, opened in January 1863, was quite a shallow affair, passing only a few feet beneath Brinnington Road.

M.S. Stokes.

104. Bredbury, 10th October 1959. This view east from the then parapet of Aston Road bridge shows Class 04 2-8-0 No 63573 with a train of mixed freight, mainly coal, initially heading for Northwich. The photographer would nowadays be risking his safety from a point redeveloped to accommodate a filling station access point and supermarket, the result of filling in the cutting to a point roughly level with the top of the loco's chimney. The tall signals protect Bredbury Junction, the signal box situated adjacent to the rear of the train. The land to the left would later be developed for industrial use along Whitefield Road but around this time however it was in use for military purposes and tanks on manouvres could frequently be seen. Following severance of the line for motorway construction a short distance towards Stockport, the line was truncated in the area of Bredbury Junction, enabling a stone terminal to be constructed. ***T Lewis***

105. Bredbury, 3rd March 1966. Keeping fit was not a problem for signallers in the Bredbury/Woodley area, especially when taking into account their task of replenishing the lamps of the tall lattice posted signals adjacent to bredbury Junction signal box. the right hand signal clears the way for this eastbound train of empty coal wagons behind A stanier 8F No 48533. The rear of the train is passing beneath Ashton Road bridge, filled in following severance of the line for motorway construction in the Brinnington area.The two running lines ahead of the train were retained to serve a stone terminal but the rail formation here has vanished beneath retail shopping developments ***G. Coltas***

106. Bredbury Junction, 1961. The apparent remoteness of Bredbury Junction signal box is falsely implied in this view from the spoil tips that separated it from the A560 road at Bredbury. The tips were a combination of coal/shale removed from the nearby long closed coal mines and provided a much used recreational facility which today's BMX enthusiasts would appreciate. Known locally as the 'Pit bonks', the materials were ultimately recovered to leave a site used initially as a golfing range and then entertainment centre. latterly, supermarket and DIY outlets have sprung up on the site. he box, an all brick structure, dated from 1931 and replaced a a tall stone built type - believed to be similar to Apethorne Junction - a few feet away on the Stockport side.It had an 18 lever frame with three spare. Following withdrawal of local passenger services over the spur to Romiley Junction at the beginning of 1967, it became obvious that that the importance of the box would diminish. That short section of line closed completely from 5th March 1967.The signal box itself remained as a Block post on the Woodley/Tiviot Dale line until 12th May 1968, by which time, the operating departments were satisfied that with the introduction of diesel traction following withdrawal of steam locomotives from Heaton Mersey MPD, the box was not needed. It was subsequently demolished during November/December 1969.The route to Romiley - seen here going off to the right - was retained because of ' complications' that may occur if the Cheadle heath to New Mills line closed. The line was eventually removed between may and July 1975 utilising an engine and train from Moston Ballast Sidings. ***G K Fox***

107. Bredbury Junction. 3rd March 1966. Situated in a shallow cutting, the trains that plied the route between Stockport and Woodley would be followed by the palls of smoke that were emitted. This view towards Woodley shows Ivatt 2-6-0 No.**43031** banking an Up freight beneath the Reddish to Romiley Junction line, only days before its withdrawal. *G. Coltas.*

108. Bredbury Junction. 3rd March 1966. The bridge seen above provides the main focal point in this view of the location where the two lines intersected. Crossing the bridge with a short Philips Park to Gowholes freight is Stanier 8F No.**48136**, of Newton Heath depot. The brake van is about to pass over a small wrought iron bridge built to carry the line over a tramway that formerly connected the mines at Woodley and Bredbury (Bents Pit). Both undertakings had ceased in the last century but earthworks remained for many years to identify the course of the tramway. Note the sign in the bottom left hand corner with the word "POINTS". This identified catch or trap points, several of which were located on the Up line to derail any errant vehicles that might decide to run away down the gradient. *G. Coltas.*

109. Bredbury. c.1965. Pollution was not a word used in the days when steam engines belched out large volumes of smoke as part of their day to day duties. Viewed from Redhouse Lane bridge at a point now almost totally transformed, this unidentified Stanier 8F makes its presence felt, the sounds reverberating against the cutting sides. A banking engine announces its arrival beneath Stockport Road bridge to the back of Bredbury station. To the left, buildings belonging to the erstwhile Bredbury Steelworks give a poignant reminder to the many families who depended on the Exors of James Mills Ltd., for their livelihood. To the right, Bredbury Station goods yard, quiet for the moment but once a thriving round the clock operation. The "bottom line", as local train enthusiasts knew it, has now been filled in and landscaped.
John Fairclough.

110. Woodley. c.1965. Another example of the tall signals to be found at this end of the Cheshire Lines system gives the right of way to a westbound coal train in the hands of an unidentified Stanier Class 5MT 2-6-0. The train has just passed beneath Mill Lane bridge, the fields to the right now occupied by housing. Once again, note the trap point on the left to check runaway vehicles. *John Fairclough.*

111. Above. Woodley junction, c.1962. The curved approach to Woodley from Tiviot Dale is nicely shown in this view from Station Road. At this point, Down trains frequently had to stop to pin down brakes. There was a 20mph speed limit but drivers were seldom over confident about going too fast. By this time the goods yard had all but closed for general merchandise but was kept busy by with coal traffic, as a Coal Concentration Depot for the district, and the scrap traffic of local trader G. Hopwood and Son who had arrived on the scene in 1955. The coal business closed from 30th December 1972, out lasting the latter by almost seven years, the site having been vacated by 6th january 1966. The route survives today as a single line connection at Woodley serving Bredbury Stone Terminal and tehe Bredbury Waste management plant. Also, the land to the right was redevoped for housing in the 1990's. ***G H Platt***

112.(above). **Woodley. c.1973.** Woodley Junction signal box outlasted most of its neighbours in the area. One of its signalmen for a number of years following closure of Bredbury Junction was "Bob" Robinson, seen here with son Douglas during a quiet moment. Bob, although not a lifelong railwayman, carried on a time honoured tradition widely associated with the industry, that of serving the local community as Chairman of the local council. He was widely known and respected in the area, retiring in 1974.
courtesy Mrs Isa Robinson.

113. Woodley. c.1962. Another view from Station Road, this time towards Hyde. The station opened in 1862 and became a junction in 1863 with the opening of the line to Stockport. The signal box was a survivor of three that originally controlled lines in the area. Woodley South, a former Midland box on the line to Romiley, was replaced sometime in the early 1930's by a ground frame. That facility was abolished in 1968 along with removal of the Up and Down loops. Woodley West, a few yards round the curve on the Glazebrook line, was a Cheshire Lines signal bx, and reduced to ground frame status about the same time as Woodley South. That ground frame was abolished in 1971, together with removal of the Down loop. The station building was originally built with two gabled sections, but extended at this end shortly after opening. The goods shed, an all timber structure, was similar in construction to those found at Marple, Strines and Hayfield, not too much of a coincidence when from an official point of view, the section was known as the Hayfield to Woodley line, bridges being numbered from Hayfield. Woodley footbridge however was the last bridge on the Hyde branch.
G.H. Platt.

114. Woodley. c.1962. This view from platform level towards Hyde shows a neat and tidy station that had remained virtually unchanged for the better part of a century. In its early days it was last stop before Portwood on the journey from Manchester London Road, a service that was to be extended to Tiviot Dale and thence to Manchester Central as the system developed. However, most of the traffic that flowed through Woodley was freight, traversing the isolated section of the CLC between Apethorne and Godley Junctions before a trip through Tiviot Dale and Glazebrook. Passenger trains were numerous, but pride of place must go to the early morning service from London, which detached some carriages for Manchester at Godley Junction before stopping at Stockport on its way to Liverpool. The author however, remembers Woodley in the 1950's when there were frequent excursions to such places as Southport, New Brighton, Chester, etc., Sunday mornings in those days brought the place to life. *G.H. Platt.*

115. Woodley. 1962. Another elevated view of the station, this time from the footpath to the Up or eastbound platform. The station still had its goods facilities, note the wagons between the shed and station buildings. At least three station staff are enjoying a break, something of a change from the days when at least thirty worked here. The footbridge, a combination of wrought and cast iron parts, echoed its Midland Railway origin although erected on a jointly run station. Originally, passengers crossed the lines by a wooden footbridge but this structure is not thought to have lasted very long. The platforms were always short and low although proposals had been put forward towards the end of the last century for their extension, chiefly for trains on the Tiviot Dale line. The cost of extending at the Hyde end proved prohibitive due to the complex nature of widening Hyde Road bridge and the scheme was never carried out. Woodley had by now settled down to life as a commuter station on the Manchester London Road to New Mills/Hayfield/Macclesfield services, although weekends brought about a rush of activity with diversions from Manchester Central during the period when the former LNWR lines out of Euston were being electrified. Nowadays, the station enjoys an hourly service, on the Manchester Piccadilly to Rose Hill route. The signal box is no more, with only a single track running in from the truncated Tiviot Dale line although this divides after a short distance, one to serve the Bredbury Stone Terminal and the other the Waste Disposal Agency. The platforms were finally raised in 1990, much to the relief of passengers. *G.H. Platt.*

116. Woodley. 1962. The view from Hyde Road bridge looking towards Apethorne Junction - just visible beyond the bridge in the distance - and Hyde. The signals in the foreground protected the station, that to the left for the Tiviot Dale line, the right for trains to Romiley. The pipe across the bottom of the picture follows the route of the original Hyde road before the coming of the railway. *G.H. Platt.*

117. Apethorne Junction. 1961. Situated between Hyde Central and Woodley, Apethorne Junction controlled movements to and from Godley Junction over a short isolated section of the CLC. Opened on 1866, this tall stone built signal box provided a landmark for over one hundred years before structural deterioration caused its demise. The fourteen lever Stevens pattern tappet frame was almost thirty feet above rail level and until abolition was open continuously. From Sunday, 12th June 1966, it ceased to be a block post, reorganization of signalling arrangements transferring control of the junction to Woodley with the points being operated by remote control. Demolition of the building took place during the months of November and December 1966, materials and spoil being loaded into rail wagons before dispersal to Guide Bridge. *W.A. Brown.*

118. Apethorne Junction. c.1965. An Ivatt Class 4 2-6-0 locomotive with its train of empty coal wagons prepares to take the Godley line. Historically, this short link relied heavily on Woodhead route traffic and with the decision in 1981 to close the electrified trans-Pennine route, it could only be a matter of time before the last trains ran between Godley and Apethorne. The gradient on this section, at 1 in 100, was not the steepest, but the speed restriction imposed on trains at this junction meant a virtual standing start. The result was impressive, with numerous pillars of smoke leaping into the sky all the way through Hyde as locomotives struggled towards the sanctuary of Brookfold where the line levelled out. *John Fairclough*

119. Tiviot Dale. 29th June 1968. The end is fast approaching for the "pretty" brick building that fronted Tiviot Dale. ***Stockport Libraries.***

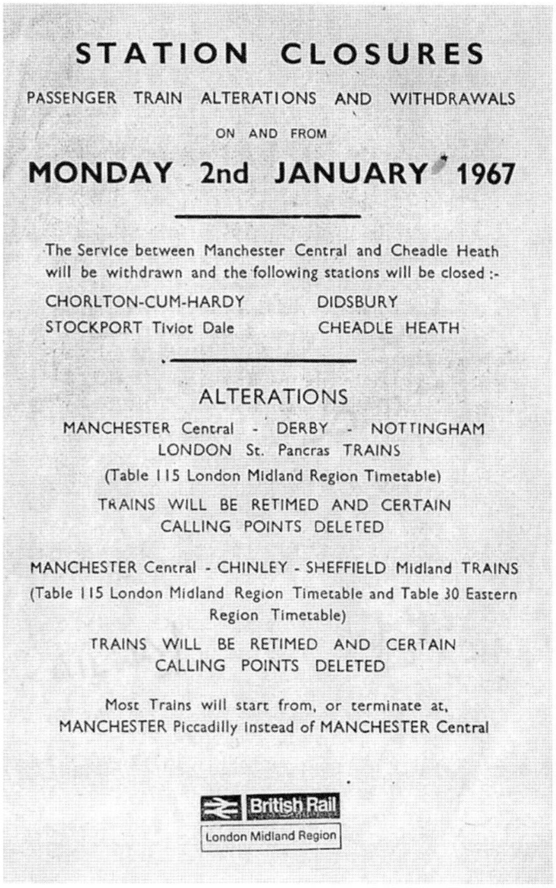

STATION CLOSURES

PASSENGER TRAIN ALTERATIONS AND WITHDRAWALS

ON AND FROM

MONDAY 2nd JANUARY 1967

The Service between Manchester Central and Cheadle Heath will be withdrawn and the following stations will be closed :-

CHORLTON-CUM-HARDY DIDSBURY
STOCKPORT Tiviot Dale CHEADLE HEATH

ALTERATIONS

MANCHESTER Central - DERBY - NOTTINGHAM
LONDON St. Pancras TRAINS
(Table 115 London Midland Region Timetable)

TRAINS WILL BE RETIMED AND CERTAIN CALLING POINTS DELETED

MANCHESTER Central - CHINLEY - SHEFFIELD Midland TRAINS
(Table 115 London Midland Region Timetable and Table 30 Eastern Region Timetable)

TRAINS WILL BE RETIMED AND CERTAIN CALLING POINTS DELETED

Most Trains will start from, or terminate at, MANCHESTER Piccadilly instead of MANCHESTER Central

British Rail
London Midland Region

120. Tiviot Dale. 17th August 1947. One is left in no doubt as to the origins of the station in this view taken in the last months of the company's independence. Although nominally operated jointly - latterly by the LMS and LNER, the station's day to day operations were carried out by numerous other sectional bodies. Signalling equipment, such as the signals themselves, and the lever locking gear, was maintained by the Signal Lineman at Stockport. Telegraph wires and associated instruments were looked after by the Telegraph Lineman, also at Stockport. Civil Engineering arrangements came under the jurisdiction of the Manchester District Officer, Mr K.C. Marrian, based in 1944 - at London Road Station, Manchester. His responsibilities and duties embraced maintenance and renewal of Permanent Way, Structures and Buildings including painting, Roads, yard surfaces, fences and drains. Gas services and water services in buildings and water mains. There was also a requirement to attend to such things as burst water pipes, leakages of gas and water, defective W.C.'s, repairs to locks and keys of drawers, buildings etc., station illuminants (oil and gas). The section of line also came under the control of the LMS Chief Mechanical and Electrical Engineers Department, Derby, which held responsibility for such things as the repairs and maintenance of weighing machines, weighbridges, locomotive water supplies. etc. Locally the work was supervised by the Outdoor Machinery Assistant based at Hunts Bank, Manchester. The district under which Stockport came was No.3, in an area extending from Padgate to Manchester Central and Godley, Chorlton, Hale to Knutsford. The LNER, not to be outdone, also had CME (Chief Mechanical Engineers Department) involvement. They were responsible, via a Carriage and Wagon Foreman at Manchester Central, for the maintenance and repair of horse-drawn vehicles, Platform barrows and trucks, Seats, Office and Waiting Room Portable Furniture, Ticket Nippers, etc, Tinware, which embraced buckets, kettles, lids, lamps etc., was repaired at Gorton Works, to which it was transported by "Lurry" from Manchester Central, the point to which items were sent by train. Shunting poles and brake sticks, of which there were numerous in the area because of the freight involvement, were dealt with by the LNER Divisional Stores Superintendent, Doncaster. It would appear that everyone had a bite at the CLC cherry, but externally, there was little evidence of all these activities. Nearly half a century later, Tiviot Dale station, along with all the other functions mentioned, has gone. *W.J. Skillern.*

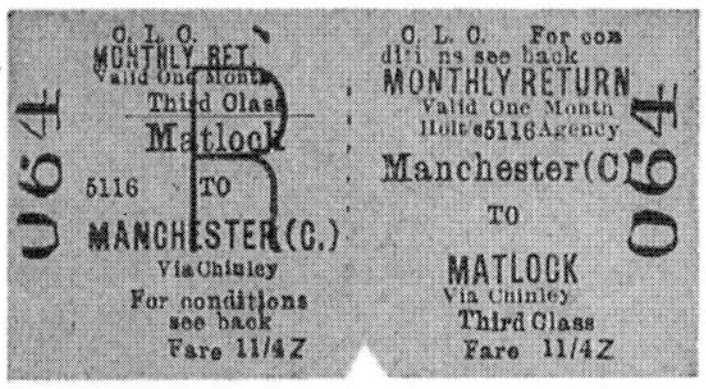

If ever there was an instance where a motorway was directly responsible for the closure of a railway, the line through Tiviot Dale was the supreme example. The alignment of the new M63 placed it within a few feet of the railway tunnel, so much so that a monitoring system was utilised to keep a check for any possible movement during construction. To ensure stability above the sandstone face, a series of rock "anchors" were installed over the top of the tunnel to a point beneath the church. It was about this time, when earthworks were still in progress, that the mobile site crane fell into a void that was present between the rock and tunnel lining. The brick lining was not penetrated, but there was sufficient concern to warrant immediate closure of the line, albeit of a temporary nature. This involved the setting up of support units comprising military trestling, a process which effectively blocked the line. A series of core holes were drilled through the brick lining to establish the extent of the void, but the conclusions did not auger well for the railway and so a temporary blockage took on an air of permanency, which resulted in re-routing arrangements for the thirty or so trains - each way - that still required the route. A press release of June 1982 commented that the Woodley - Northenden Junction freight line had been closed for over two years for safety reasons. It was not the intention at the time to close it permanently but with subsequent changing freight patterns, the question of closure was to be studied. The rest, as they say, is history. The three views on this page show the transformation from a working railway, albeit in decline, to the ultimate in dereliction, a disused formation. The top photograph (**121**) was taken in 1965 and shows a B.R. Standard Class 9F 2-10-0 hauling a train of empty coal wagons, possibly on their way back to Yorkshire.The centre view (**122**), recorded towards the end of 1968, shows the station following demolition of the main buildings and removal of platform lines. This illustrates the negative side of transport thinking, for in all probability, the eventual cost of derelict land reclamation may well be more than was saved by withdrawing the services in the first place. The third and final view (**123**) was taken in 1986. What more is there to say ...**the end**.

A. Barnard.